Behaviour of Energetic Coherent Structures in Turbulent Pipe Flow at High Reynolds Numbers

Behaviour of Energetic Coherent Structures in Turbulent Pipe Flow at High Reynolds Numbers

Von der Fakultät für Maschinenbau, Elektro- und Energiesysteme der Brandenburgischen Technischen Universität Cottbus-Senftenberg zur Erlangung des akademischen Grades eines

Doktor der Ingenieurwissenschaften (Dr.-Ing.)

genehmigte Dissertation
vorgelegt von

M.Sc.
Zeinab Hallol

geboren am 18.12.1978 in Kairo

Vorsitzender: Prof. Dr. rer. nat. Andreas Schröder (DLR)
Gutachter: Prof. Dr.-Ing. Christoph Egbers (BTU)
Gutachter: PD Dr. Olga Shishkina (Max Planck Institute)

Tag der mündlichen Prüfung: 10.09.2021

Bibliografische Information der Deutschen Nationalbibliothek
Die Deutsche Nationalbibliothek verzeichnet diese Publikation in der
Deutschen Nationalbibliografie; detaillierte bibliographische Daten sind im Internet
über http://dnb.d-nb.de abrufbar.
1. Aufl. - Göttingen: Cuvillier, 2021
Zugl.: Brandenburgische Technische Universität Cottbus-Senftenberg, Diss., 2021

Nonnenstieg 8, 37075 Göttingen
Telefon: 0551-54724-0
Telefax: 0551-54724-21
www.cuvillier.de

1. Auflage, 2021
Gedruckt auf umweltfreundlichem, säurefreiem Papier aus nachhaltiger Forstwirtschaft.

ISBN 978-3-7369-7501-9
eISBN 978-3-7369-6501-0

Abstract

In this thesis, coherent turbulent structures (large and very large-scale motions) in turbulent pipe flow are investigated at relatively high Reynolds numbers and study their association in both total kinetic energy and Reynolds shear stress. Experimental investigations have been performed in CoLaPipe (Cottbus-Large-Pipe) for pipe flow over a wide range of Reynolds number $8 \times 10^4 \leq Re_D = u_b D/\nu \leq 1 \times 10^6$, where u_b is the bulk velocity and D is pipe diameter, located at the Aerodynamics and Fluid Mechanics Department, Brandenburg University of Technology Cottbus- Senftenberg (BTU).

The first part of the thesis focuses on determining the contribution of the coherent structures using one-dimensional spectral analysis and assessing the structures behaviour in the outer region of pipe flow using high spatial resolution Hot-wire measurement up to 30kHz. The results of the power and pre-multiplied spectrum of stream-wise velocity indicate that the wavelength value of very large scale motions (VLSMs) acquires 19R at a maximum Reynolds number range $Re_D = 1 \times 10^6$(Re_τ=19000). On the other hand, large-scale motions (LSMs) have a wavelength value of 3R over different Reynolds number range.

Regarding the identified wavelength values, it is observed that contribution to energy for structures greater than 3R carries 55 % of total kinetic energy. In addition, temporal-spatial resolution using the High-speed PIV measurements with 1838 snapshots has been performed in CoLaPipe to estimate the contribution magnitude of stream-wise/wall-normal velocity fluctuations to total kinetic energy and Reynolds shear stress in the logarithmic and outer layer. The stream-wise contour spectra indicated that at Reynolds number Re_τ= 3200 based on friction velocity, signature of foot print of large structures with wavelength value (λ_x) $<$ 3R are apparent in the near-wall region, while very large scale motions (VLSMs) $\lambda_x >$ 3R are manifest in the outer layer at all investigated Reynolds numbers. The co-spectra illustrates that the contribution of VLSMs to the Reynolds shear stress is about 45% to 60%. This contribution is approximately equivalent to the small and large scale structures association in the near-wall region.

Furthermore, Energy contributions in the POD modes are explored to understand the spatial-temporal characteristics of the coherent structures. The first three POD modes are recognised as the most essential in turbulent kinetic energy contribution. However, large scale structures are considered the most energetic to the first mode with respect to the other POD modes (Wu 2014). It indicated that the large-scale fluctuating velocity fields contribute strongly through u velocity component (u^{2+}) to carry large amounts of kinetic energy while no contribution from v component (v^{2+}) is observed. Consequently, high contribution of Q_2 / Q_4 events is demonstrated to Reynolds shear stress.

Acknowledgements

Firstly, I would like to express my sincere gratitude to my supervisor Prof. Dr Ing. Christoph Egbers, for the continuous support during PhD study and related research, for his patience, motivation, and immense knowledge. His guidance helped me in all the time of research and writing of this thesis. I could not have imagined having a better supervisor and mentor for my PhD study.

Besides my supervisor, I would like to thank the rest of my thesis committee: Prof. Shishkina, and Prof. Schröder, for their insightful comments, questions and encouragement.

I would also like to thank my tutor, Dr Ing. Sebastian Merbold, for his valuable guidance throughout my studies. You provided me with the tools that I needed to choose the right direction and successfully completed my dissertation.

I would like to thank Frau Hendrischke and her colleagues in the central equality office (Zentrale Gleichstellung) at the Brandenburg University of Technology that has financially supported my PhD research. Without their support, this dissertation was not complete in a successful way. To my working colleagues in the Aerodynamics and Fluid Mechanics chair; Prof. Harlander, Frau Kaschwich, Herr. Stöckert, Dr Vasyl, Jurg, Ludwig and Mohamed Yousry, and other colleagues for their support and help in the last four years.

I would particularly like to thank my former supervisor Prof. Dr Mostafa Gouda, as you were the first one who believes in me and provides me with advice, support and encourages me to proceed with my research in Egypt.

Last but not least, I would like to thank my Mother for her wise counsel and sympathetic ear. You are always there for me. I also would like to thank my Father. I will be ever grateful for his assistance and love, and I am sorry that he has not lived to see my achievement. Finally, I could not have completed this dissertation without the support of my husband and my children, who were supporting me spiritually throughout writing this thesis and my life in general.

Declaration of Authorship

I, Zeinab Hallol, declare that this thesis titled, Behaviour of Coherent Structures in Turbulent Pipe Flow at High Reynolds Numbers and the work presented in it are my own. I confirm that:

- This work was done wholly or mainly while in candidature for a research degree at this University.
- Where any part of this thesis has previously been submitted for a degree or any other qualification at this University or any other institution, this has been clearly stated.
- Where I have consulted the published work of others, this is always clearly attributed.
- Where I have quoted from the work of others, the source is always given. With the exception of such quotations, this thesis is entirely my own work.
- I have acknowledged all main sources of help.
- Where the thesis is based on work done by myself jointly with others, I have made clear exactly what was done by others and what I have contributed myself.

You will never gain knowledge without [possessing]
six [qualities],
Sharpeness [of the mind]
eagerness [to learn]
sacrifice [in terms of time etc]
means [i.e wealth]
the company of a scholar
and length of time!

Imaam Shafi'ee

Contents

Contents viii

List of Figures xi

List of Tables xvi

List of Symbols xvii

1 Introduction 1

1.1 Motivation . . . 1

1.1.1 How is the High Reynolds Number Differ from the Low Reynolds Number? . . . 2

1.1.2 Turbulent Structures . . . 5

1.1.3 Pipe Flow Boundary Layers . . . 8

1.2 Theoretical Background . . . 10

1.2.1 Pipe Flow . . . 11

1.3 Objectives of the Thesis . . . 12

1.4 Outline of the Thesis . . . 13

2 Experimental Facilities and Measurements Techniques 15

2.1 CoLaPipe . . . 17

2.2 Measurement Techniques . . . 21

2.2.1 Hot-Wire Anemometer . . . 21

2.2.2 Types of Hot-Wire Probes . . . 22

2.2.3 Overheat Ratio . . . 23

2.2.4 Hot-Wire Anemometer Velocity Calibration . . . 24

2.2.5 Polynomial Curve Fitting . . . 24

2.2.6 Power Law Curve Fitting . . . 25

2.2.7 Particle Image Velocimetry (PIV) 26
2.2.8 PIV Main Components 27
2.2.9 How Nd: YLF Laser Emits the Laser Ray ? 27
2.2.10 Pressure Gradient Measurements 29
2.2.11 Application at CoLaPipe Test Facility 29
2.2.12 Fluid Parameters 30

3 Influence of Calibration Methods on HWA Measurements 32
3.1 Theoretical Background 32
3.2 Experimental Setup 33
3.3 Results 36
3.4 Conclusion 44

4 Kinetic Energy Contribution of Coherent Structures in Fully Developed Turbulent Pipe Flow at High Reynolds Numbers 45
4.1 Experimental Background 45
4.2 Experimental Setup 47
4.3 Validation of Hot-Wire Measurements 48
4.4 Results 50
4.5 Conclusion 62

5 PIV Measurements in Pipe Flow 64
5.1 Introduction 64
5.2 Experimental Setup 64
5.3 PIV Measurements Validation 66
5.3.1 Validation of Spectral Analysis 68
5.4 Characterization of Turbulent Structures 70
5.5 Temporal-spatial Analysis 75
5.6 Quadrant Analysis 77

6 Two-dimensional Spectral Analysis in Pipe Flow 78
6.1 Introduction 78
6.2 Results 80
6.2.1 Pre-multiplied Spectral Analysis 80
6.2.2 Contribution of Reynolds Shear Stress 83
6.3 Conclusion 87

7 Proper Orthogonal Decomposition Analysis (POD) 89
7.1 Introduction 89
7.2 Proper Orthogonal Decomposition 90
7.3 POD Analysis 92
7.3.1 Contributions to Reynolds Stresses ($-uv^+$, u^{2+}, and v^{2+}) 97
7.3.2 Q_2 and Q_4 Events Contribution in Reynolds Stresses 99
7.4 Conclusion 100

8 Conclusions and Final Remarks 102
8.1 Comparison Between In-situ and Ex-situ Calibration Methods 102
8.2 One-dimensional Spectral Analysis 103
8.3 Two-dimensional Spectral Analysis 104
8.4 Proper Orthogonal Decomposition Analysis 106
8.5 Recommendations for Future Work 107

Bibliography 109

List of Figures

1.1 Turbulence kinetic energy production for a range of Reynolds numbers: (a) Semi-logarithmic representation and (b) Pre-multiplied representation (where equal areas represent equal contributions to the total production). Adopted from Marusic et al. 2010. 3

1.2 Contour maps showing the variation of one-dimensional pre-multiplied spectra with wall-normal position for two Reynolds numbers. An inner and an outer peak are noted at the higher Reynolds number. Adopted from Hutchins and Marusic 2007a. 4

1.3 Flow Structures in turbulent boundary layers. 6

1.4 Conceptual model which describes the alignment of hairpins coherently into a package to form VLSM. (Kim and Adrian 1999) 7

1.5 Schematic sketch of various regions and layers of pipe and channel flows, adopted from Wosnik et al. 2000. 9

2.1 Reynolds famous experimental setup (adopted from Rott 1990). 16

2.2 Comparsion between different pipe test facilites regarding aspect ratio (L/D), internal diameter and maximum Reynolds number range. (Öngüner 2018) 17

2.3 Sketch of the CoLaPipe at the Department of Aerodynamics and Fluid Mechanics (LAS BTU Cottbus-Senftenberg). (König 2015) 19

2.4 The 28 m long CoLaPipe facility in LAS. 19

2.5 Development of statistical quantities along pipe test section at $Re_D = 1.8 \times 10^5$ (open symbols) and $Re_D = 4 \times 10^5$ (filled symbols). (König 2015) . 20

2.6 Wall pressure fluctuations at $Re_D \approx 300.000$ along the CoLapPipe for different ring disturbances (Selvam et al. 2018) 20

2.7 Hot-wire anometery probes.(a) Single straight hot-wire probe (55P11), and (b) Boundary layer hot-wire probe (55P15). Dantec website. 22

2.8 (a) Relation between sensor resistance and temperature, and (b) Wheatstone bridge circuit. Adopted from Dantec website. 23
2.9 Hot-wire anemometry calibration curve. 25
2.10 Schematic drawing of the PIV set up. Adopted from Chanetz et al. 2020. 26
2.11 Schematic drawing for the Nd:YLF crystal. Raffel et al. 2007 28
2.12 Photograph of pipe test section provided with pressure tapping for measurement of static pressure along the pipe. 29

3.1 Sketch of the HWA experimental setup in CoLaPipe test facility. (König 2015) 33
3.2 Left:Ex-situ calibration unit with air supply.Right:Calibration unit contraction with probe to be calibrated. 35
3.3 Calibration profiles for a single wire probe. o; In-situ calibration data points, x; Ex-situ calibration data points, –; 4th order polynomial. 36
3.4 Logarithmic velocity profiles for In-situ and Ex-situ calibration methods. (a) $Re_\tau = 6600$, and (b) $Re_\tau = 13000$. 37
3.5 (a) Root mean square (RMS) profiles for In-situ and Ex-situ calibration methods at $Re_\tau = 6600$, and (b) RMS profiles for In-situ and Ex-situ calibration methods at $Re_\tau = 13000$. 38
3.6 Turbulent Intensity velocity profile (u'^2/u_τ^2) for the In-situ and Ex-situ datasets at $Re_\tau = 6600$ and 13000 compared to the data from Zagarola and Smits 1998. 39
3.7 Kurtosis profiles for In-situ and Ex-situ calibration methods versus DNS data $Re_\tau = 1080$ (Penga et al. 2018). (a)$Re_\tau = 6600$, and (b) $Re_\tau = 13000$. 40
3.8 Skewness profiles for In-situ and Ex-situ calibration methods versus DNS data $Re_\tau = 1080$ (Penga et al. 2018). (a)$Re_\tau = 6600$, and (b) $Re_\tau = 13000$. 40
3.9 Spectra profiles for the In-situ and Ex-situ calibration methods at $Re_\tau 6600$ (a) y/R=0.12, (b) y/R=0.3, (c) y/R=0.76, and (d) y/R=1. 41
3.10 Spectra profiles for the In-situ and Ex-situ calibration methods at $Re_\tau 13000$ (a) y/R=0.12, (b) y/R=0.3, (c) y/R=0.76, and (d) y/R=1. 42
3.11 Pre-multiplied spectra contours for the In-situ and Ex-situ calibration methods at different Reynolds number. Black and red symbol (x) represents the outer peak location for the In-situ and Ex-situ respectively. 43

4.1 Statistical moments of the stream-wise velocity component for the present experimental data compared to experimental and numerical datasets. (a) Mean velocity, (b) Stream-wise Reynolds stress, (c) Skewness, and (d) Kurtosis. 50

4.2 Power spectrum of stream-wise velocity fluctuation scaled by pipe radius (R). (a) $Re_\tau = 2156$, (b) $Re_\tau = 6556$, (c) $Re_\tau = 13132$, and (d) Re_τ=19000. 52

4.3 Power spectrum of stream-wise velocity fluctuation scaled by wall-normal locations (y). (a) $Re_\tau = 2156$, (b) $Re_\tau = 6556$, (c) $Re_\tau = 13132$, and (d) Re_τ=19000. 53

4.4 Pre-multiplied power spectrum of stream-wise velocity fluctuation (Φ_{uu}/u_τ^2) versus wavenumber scaled by pipe radius $(k_x R)$.(a) $Re_\tau = 2156$, (b) $Re_\tau = 6556$, (c) $Re_\tau = 13132$, and (d) Re_τ=19000. 54

4.5 Pre-multiplied power spectrum of stream-wise velocity fluctuation (Φ_{uu}/u_τ^2) versus wavenumber scaled by wall-normal locations $(k_x y)$. (a) $Re_\tau = 2156$, (b) $Re_\tau = 6556$, (c) $Re_\tau = 13132$, and (d) Re_τ=19000. 55

4.6 Upper row: Contour plots of spectra $(k_x \Phi_{uu}/u_\tau^2)$. Bottom row: turbulent intensity (u^{2+}) profiles. (a) $Re_\tau = 2156$, (b) $Re_\tau = 6556$, (c) $Re_\tau = 13132$, and (d) Re_τ=19000. Symbol (X): location of the outer spectral peak y^+_{osp}. 57

4.7 Contour plots of spectra $(k_x \Phi_{uu}/u_\tau^2)$ in outer scaling. (a) $Re_\tau = 2156$, (b) $Re_\tau = 6556$, (c) $Re_\tau = 13132$, and (d) Re_τ=19000. 58

4.8 Wavelengths of the peaks in the pre-multiplied power spectra of stream-wise velocity. 59

4.9 Cumulative stream-wise kinetic energy fraction (γ_{uu}) for $0.048 \leq$ y/R ≤ 0.28. (a) Re_τ= 2156,(b) Re_τ= 6556, (c) Re_τ= 13132, and (d) Re_τ= 19000.Vertical dashed lines refer to the position of $\lambda_x/R = 3$. 60

4.10 Cumulative stream-wise kinetic energy fraction (γ_{uu}) for $0.28 \leq$ y/R ≤ 0.7. (a) Re_τ= 2156,(b) Re_τ= 6556, (c) Re_τ= 13132, and (d) Re_τ= 19000.Vertical dashed lines refer to the position of λ_x/R=3. 61

5.1 PIV setup in CoLaPipe test facility at LAS. 66

5.2 Comparison between the present PIV experimental datasets versus DNS datasets at $Re_\tau = 3008$ (Ahn et al. 2015), LES dataset $Re_\tau = 1000$ (Chin et al. 2015), $Re_\tau = 10480$ (Zagarola and Smits 1998), $Re_\tau = 5080$ (Morrison et al. 2004), and $Re_\tau = 6556$ (HWA measurements). (a) Logarithmic velocity profile (u^+), (b) Stream-wise and wall-normal turbulent Intensity (u^{2+} and v^{2+}) and Reynolds Shear stress ($-uv^+$), (c) Skewness, and (d) Kurtosis. 67

5.3 Wavenumber spectrum velocity fluctuation for the present PIV datasets ($Re_\tau = 3200$) versus other experimental pipe flow ($Re_\tau = 3815$) (Guala et al. 2006) at y/R=0.3. (a) Power spectra for stream-wise velocity fluctuation, and (b) Power spectra for wall-normal velocity fluctuation. . 69

5.4 Contour plots of mean stream-wise velocity (u_m) normalised by corresponding mean velocity at pipe axis (u_c). 71

5.5 Contour plots of stream-wise (u_{rms}) velocity normalised by friction velocity (u_τ). 72

5.6 Contour plots of Reynolds shear stress velocity (uv) normalised by friction velocity (u_τ). 73

5.7 Sequential snapshots of PIV instantaneous fluctuations velocity fields with equivalent time interval between each other Δ t=0.002s at Re_τ= 3200. . 74

5.8 Temporal-spatial of the stream-wise fluctuation velocity u' (left) and Reynolds shear stress $u'v'$ (right) for the PIV dataset Re_τ= 1667 at different wall-normal locations. (a) y/R= 0.1, (b) y/R= 0.3, (c) y/R =0.5, and (d) y/R = 0.7. 76

5.9 Temporal-spatial of the stream-wise fluctuation velocity u' (left) and Reynolds shear stress $u'v'$ (right) for the PIV dataset Re_τ= 3200 at different wall-normal locations. (a) y/R= 0.1, (b) y/R= 0.3, (c) y/R =0.5, and (d) y/R = 0.7. 76

5.10 Quadrant analysis plots for the fluctuation velocity components (u' and v') at Re_τ=10617. (a) y/R= 0.1, and (b) y/R= 0.8. 77

6.1 Upper row: Contour plots of spectra ($k_x\Phi_{uu}/u_\tau^2$) in inner scale . Bottom row: Turbulent intensity (u^{2+}) profiles. (a) $Re_\tau = 3200$, (b) $Re_\tau = 6605$, and (c) Re_τ= 10617. Symbol (X): location of the outer spectral peak y^+_{osp}. 81

6.2 Contour plots of spectra ($k_x\Phi_{uu}/u_\tau^2$) in outer scale. 82

6.3 Fraction kinetic energy (γ_{uu}) carried out by VLSMs along the wall-normal locations. 83

6.4 Contour plots of co-spectra ($k_x\Phi_{uv}/u_\tau^2$) in inner scale. 84

6.5 Contour plots of co-spectra ($k_x\Phi_{uv}/u_\tau^2$) in outer scale. 85

6.6 Fraction Reynolds shear stress (γ_{uv}) carried out by VLSMs along the wall-normal locations. 86

7.1 Kinetic energy content and cumulative energy of the first 20 POD modes. 93

7.2 POD spatial modes of the velocity fields (a) First mode, (b) Second mode, and (c) Third mode at Re_τ = 10617. 94

7.3 (a) Scatter plot of the first POD time coefficient (a_1) normalized by its RMS value for Re_τ=10617, and (b) Histogram of normalized a_1. 95

7.4 An instantaneous fluctuating velocity field with a large positive POD coefficient of $a_1 = 2.17\sigma_{a1}$ for the first POD mode. 96

7.5 An instantaneous fluctuating velocity field with a large negative POD coefficient of a_1 = -2.17σ_{a1} for the first POD mode. 97

7.6 Comparisons of Reynolds stresses between the original ensemble of the fluctuating velocity fields and the ensemble without those velocity fields whose $\mid a_1 \mid > 2\sigma_{a1}$. (a) Stream-wise Reynolds stress (u^2); (b) Wall-normal Reynolds stress (v^2); and (c) Reynolds shear stress ($-uv$). 98

7.7 Comparisons of Reynolds stresses between the original ensemble of the fluctuating velocity fields and the ensemble without those velocity fields. (a) Stream-wise Reynolds stress (u^2); and (b) Reynolds shear stress ($-uv$). 99

8.1 Comparsion between the pre-multiplied spectra of stream-wise velocity fluctuation obtained using HWA and PIV datasets at the same Reynolds number Re_τ =6556. 105

8.2 Pre-multiplied contour spectra (inner scaled) of stream-wise velocity fluctuation for the two implemented measurement technique, HWA and PIV datasets at Re_τ =6556. 105

List of Tables

2.1 Various pipe test facilities. 17

4.1 Experimental flow parameters: Reynolds number (Re_D) based on diameter, Reynolds number (Re_τ) based on friction velocity, centreline (u_c), bulk (u_b), friction (u_τ) velocities, density (ρ), and kinematic viscosity (ν) of the fluid. 48

5.1 Experimental flow parameters. Re_D is Reynolds number based on pipe diameter and bulk velocity (u_b), while Re_τ is Reynolds number based on pipe radii and friction velocity (u_τ) 65

List of Symbols

Latin Symbols

a_n	POD temporal coefficient
a_w	Overheating ratio
CoLaPipe	Cottbus large pipe test facility
C_R	Contraction ratio
D	Pipe diameter
d_w	Hot-wire probe diameter
dp/dx	Pressure gradient
E	Hot-wire probe voltage
Ex-situ	Off site (external) calibration
In-situ	In site (internal) calibration
f	Frequency
HWA	Hot-wire anemometry
HS-PIV	High-speed particle image velocimetry
k_x	Stream-wise wavenumber
L	Pipe length
l_c	Wall turbulence length scale
l_w	Hot-wire probe length
LDA	Laser doppler anemometry
PIV	Particle image velocimetry
p	Turbulence kinetic energy production
p	Total pressure
P_{uu}	Power spectra in frequency domain
R	Pipe radius
$\mathcal{R}$	Universal gas constant

$\mathbb{T}$	Flow temperature
T_{amb}	Ambient temperature
T_w	Operating temperature
T	Time domain
t_c	Convective time scale
Δt	Time variance
TKE	Turbulent kinetic energy
u	Stream-wise velocity components
v	Wall-normal velocity components
w	Span-wise velocity components
u_τ	Friction velocity
u_b	Bulk velocity
u_c	Centreline velocity
u_m	Mean velocity
u_{rms}	Root mean square velocity
u'	Fluctuating velocity
$u_{recal.}$	Recalculated velocity
u_{co}	Convective velocity
$-u'v'$	Reynolds shear stress
u_∞	Free stream velocity
LSMs	Large scale motions
VLSMs	Very large scale motions
x	Stream-wise flow coordinate
y	Wall-normal flow coordinate
z	Span-wise coordinate

Dimensionless numbers

R^+	Kármán number
Re_τ	Reynolds number based on friction velocity and pipe radii
Re_D	Reynolds number based on bulk velocity and pipe diameter
p'^+	Wall pressure fluctuation
y^+	Dimensionless wall distance
y^+_{osp}	Outer spectral peak location
λ^+	Normalized viscous scaled stream-wise wavelength

l^+	Hot-wire probe viscous scaled length
u^+	Normalized stream-wise velocity
u^{2+}	Normalized stream-wise turbulent Intensity
v^{2+}	Normalized wall-normal turbulent Intensity

Greek letters

δ	Boundary layer thickness
ΔT	Temperature difference
τ	Shear stress
τ_w	Wall shear stress
Φ_{uu}	Stream-wise power spectra in wavenumber space
Φ_{uv}	Power co-spectra of the stream-wise, u, and wall-normal, v, velocity
Φ	POD spatial domain
k_x	Stream-wise wavenumber
λ	POD eigenvalue
λ_x	Stream-wise wavelength
λ_{max}	Maximum wavelength scales
μ	Dynamic viscosity
ν	Kinematic viscosity
ρ	Air density
ϵ	Dissipation scale
α	Asymptotic value of the fluctuation pressure
γ_{uu}	Cumulative stream-wise kinetic energy fraction
γ_{uv}	Cumulative Reynolds shear stress fraction
σ_{a1}	Root mean square value of the temporal coefficient

Dedicated to my father's soul

Chapter 1: Introduction

1.1 Motivation

Turbulent flows exist in nature, and many engineering applications encountered in our daily life, such as water streams in rivers, the motion of the air in the earth atmosphere, blood in our veins, and fluids in pipes ... etc. Turbulence is the most complicated and challenging form of fluid motions. For that, Tennekes and Lumley 1972 defined turbulence as a comprehensive description instead of giving a precise definition with the main characterisation represented in; Large Reynolds number, diffusivity, desperation at small scales, three-dimensional vorticity fluctuations, and irregularity. In the study of turbulence, devising methods to separate the complex turbulent motions into simplified events called (coherent structures) has been a challenging task. However, by increasing the Reynolds number, the separation in scales increases and analysing spatially and temporally. In wall-bounded turbulent flows, a grasped the attention for intensive research based on experimental and numerical analysis had been proceeded to understand the mechanism of coherent structures. Investigation of coherent structures in turbulent boundary layers has caught great experimental and numerical interest in the last decades, as they are known to play a significant role in the production of turbulent kinetic energy (Ganapathisubramani et al. 2005 and Vallikivi et al. 2015). In spite of the occurrence and length of these structures that have been documented, the relationship between these coherent structures and the energy spectrum has not yet been fully resolved (Srinath 2017) .

Although, researchers start to implement direct numerical simulation (DNS) due to its known advantages, such as resolving all relevant turbulent length and time scales. Nevertheless, its computational power is incapable of investigating at high Reynolds numbers and complex geometries. A recent effort was done to implement a new numerical technique, so-called characteristic dynamic mode decomposition (CDMD). It was performed to approximate the coherent structures head vortex as dynamic modes in space and time (Sesterhenn and Shahirpour 2019).

In the foreseeable future, direct numerical simulation is still far beyond the needs of the investigations and limited to very low Reynolds numbers and simple shapes. For that, direct numerical simulation is not a realistic possibility in most cases compared to the experimental investigation, and in any event, simulation by itself does not substitute the need for experimental investigations.

In addition, due to the lack of sufficiently high Reynolds number data and large test facilities, understanding of wall turbulence through experimental investigations are sparse. However, the length of large-scale structures up to 14 δ in external flow (Kim and Adrian 1999; Hutchins and Marusic 2007a; Lee and Sung 2013) and needs a large field of view and high spatial resolution. SuperPipe in Princeton University and Center for International Cooperation in Long Pipe test facility (CICLoPE) in Bolognia University are the most popular large test facilities used to investigate turbulent pipe flow structures. In the Aerodynamics and Fluid mechanics department, Cottbus Large Pipe test facility (CoLaPipe) was built to improve the understanding of turbulent flow by performing the characterisation of the flow (Drag, flow structures, boundary layers, ...etc.).

1.1.1 How is the High Reynolds Number Differ from the Low Reynolds Number?

Most of the previous experimental investigations in the wall turbulence field have been performed at low and moderate Reynolds numbers. This range of Reynolds number was preferred as it provides a physically thick viscous layer at the near-wall region. In addition, researchers were driven by the fact that the peak kinetic energy occurs at the viscous buffer region where $y+ \approx 12$. This fact occurs at a low Reynolds number where the major contribution to bulk production occurs at the near-wall region. However, the major contribution to the turbulence at high Reynolds numbers produced at the logarithmic region appears in fig.(1.1) as the logarithmic region dominates at sufficiently high Reynolds numbers. It was reported that the contribution of the bulk production to the logarithmic region was equal to the near-wall region at $y^+ \leq 30$ at $Re_\tau = 4200$ (Smits et al. 2011).

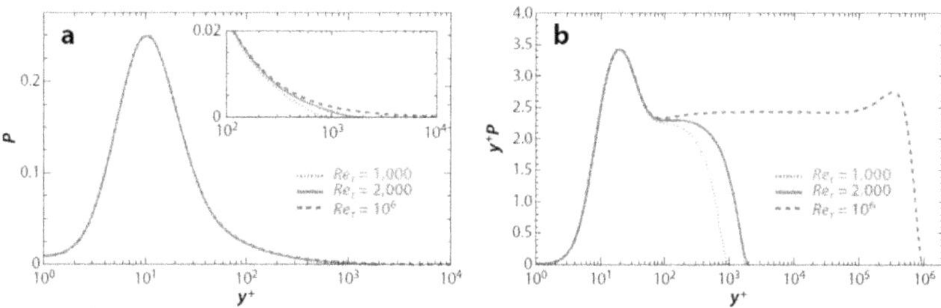

Figure 1.1: Turbulence kinetic energy production for a range of Reynolds numbers: (a) Semi-logarithmic representation and (b) Pre-multiplied representation (where equal areas represent equal contributions to the total production). Adopted from Marusic et al. 2010.

Furthermore, experimental investigations show that in pipe flow, the logarithmic region starts for $y^+ \leq 600$ (Zagarola and Smits 1998). However, they mentioned that the logarithmic variation was observed at $Re_\tau > 50000$. Normally, the classical estimation was known for the logarithmic region for wall-normal range 30 ν/u_τ < y < 0.15δ need only Re_τ= 2000. Similar behaviour was reported in the boundary layer but at higher friction Reynolds number Re_τ = 13300 (Nagib et al. 2007; Sreenivasan and Sahay 1997). Another advantage of the high Reynolds numbers experimental investigation was reported by Hutchins and Marusic 2007a,b. They investigate the stream-wise velocity spectrogram in the boundary layer flow at low and relative high Reynolds number. As depicted in fig.(1.2), they clearly observe the two distinct peaks in the stream-wise velocity pre-multiplied spectra at higher Reynolds number as the very large scale peak was hard to distinguish at the low Reynolds numbers. Hutchins and Marusic 2007b proposed that the spectral peaks need $Re_\tau > 4000$ is required to ensure the occurrence of the scale separation of high Reynolds number turbulence.

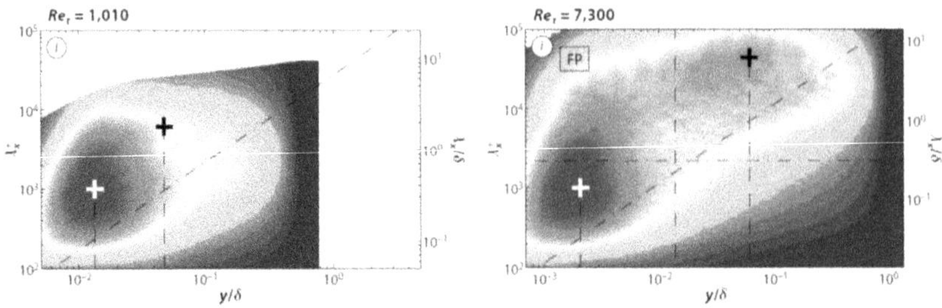

Figure 1.2: Contour maps showing the variation of one-dimensional pre-multiplied spectra with wall-normal position for two Reynolds numbers. An inner and an outer peak are noted at the higher Reynolds number. Adopted from Hutchins and Marusic 2007a.

Turbulent coherent structures, large scale motions (LSMs) and very large scale motions (VLSMs), are studied inside the first phase of the Schwerpunktprogramme (SPP) at the aerodynamics and fluid mechanics department of Brandenburg University of Technology (BTU) experimentally in turbulent pipe flow at high Reynolds numbers using hot-wire anemometry measurement technique. This program SPP-1881 was funded by Deutsche Forschungsgemeinschaft (DFG). The results of the SPP first phase are reported, such as one-dimensional spectral analysis conducted in CICoLPE (Öngüner et al. 2017a) and CoLaPipe (Zanoun et al. 2017) test facilities at low and high Reynolds numbers. Preliminary PIV measurement (100 snapshots) at CoLaPipe was used in applying auto-correlation analysis (Öngüner et al. 2017b). In addition, Zanoun et al. 2019 examined the direct and indirect measurement techniques for measuring the Reynolds stress tensor in low and high Reynolds numbers.

The objective of the second phase is to characterize further, experimentally and numerically, the large-scale turbulent structures (LSMs and VLSMs) in turbulent pipe flow at low and high Reynolds numbers. In support of the SPP second phase, the present dissertation is focused on studying the coherent structures in logarithmic and outer region experimentally at relatively high Reynolds numbers. The experiments are carried out by using high-speed PIV and HWA at high spatial and temporal resolution. The main aim of this thesis is to represent the answers to the following questions:

- Does the length scale value of large scale motions and very large scale motions provide a significant change by increasing Reynolds number?

- To which extend does the wavelength of large and very large scale motions persists in the pipe flow?
- How LSMs and VLSMs contribute to stream-wise turbulent kinetic energy and Reynolds shear stress at high Reynolds numbers in pipe flow?
- What is the contribution of large scale structures in the first proper orthogonal decomposition (POD) mode?

1.1.2 Turbulent Structures

Turbulent flows are known by the characteristic recurrent forms of structural packets. These structured packets are collectively known as coherent structures (Holmes et al. 2012). Coherent structures in canonical wall-bounded turbulent flows are commonly used to interpret and understand turbulent physics (Theodorsen 1952; Robinson 1991; Jiménez and Moin 1991; Adrian et al. 2000). Many of the structures that have been identified have been common among the three canonical wall-bounded flows (boundary layers, channels, and pipes) as shown in fig.(1.3). It is observed from studies that these structures are energetically dominant in many flows (Perry and Marusic 1995; Marusic et al. 2010; Monty et al. 2009; Mckeon 2017). Turbulent structures and their contribution to turbulent statistics came to prominence decades ago (Adrian 2007). The bulk motions of large-scale structures with stream-wise extent $\sim$ o(R), where R is the pipe radius, can be divided into large and very-large-scale motions (LSMs and VLSMs) (Adrian et al. 2000; Kim and Adrian 1999). The large-scale motions (LSMs) are recognised to be created by the vortex packets formed when multiple hairpin structures travel at the same convective velocity as shown in fig.(1.4) (Kim and Adrian 1999; Zhou et al. 1999; Guala et al. 2006; Balakumar and Adrian 2007). LSMs have characteristic features representing in the hairpin vortices within the packet align in the stream-wise direction and induce regions of low stream-wise momentum between their legs (Brown and Thomas 1977; Adrain et al. 2000; Ganapathisubramani et al. 2003; Tomkins and Adrian 2003; Hutchins et al. 2005). Its length scale is demonstrated to be common of approximately 2-3R for pipe and boundary layer in stream-wise direction. On the other side, a structural element of wall-bounded turbulent flow that has recently received attention is the "regime of very long meandering positive and negative stream-wise velocity fluctuations" (Hutchins and Marusic 2007a). These structures are defined as superstructures but are also commonly known as very large-scale motions (VLSMs) in internal flows (e.g. pipes and channels) and superstructures in external flow (e.g. boundary layers).

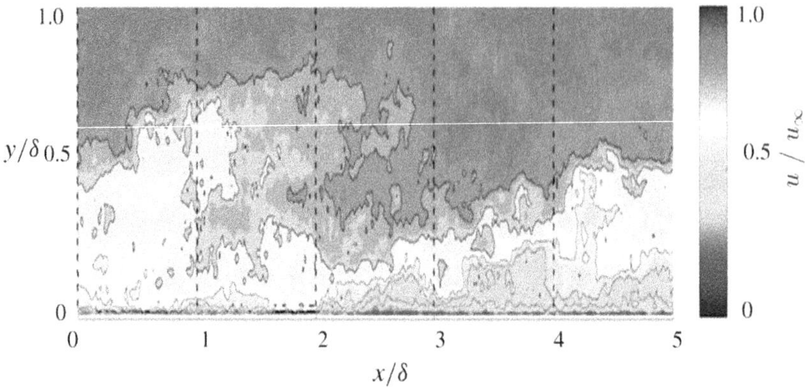

(a) Instantaneous stream-wise flow field for boundary layer. (Saxton and Mckeon 2017)

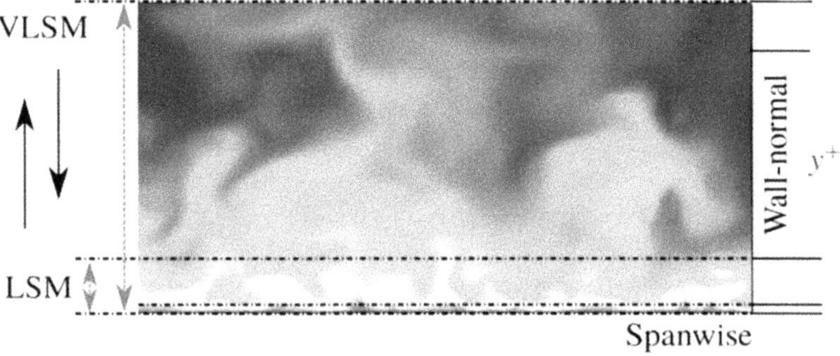

(b) Stream-wise velocity contour in the cross-stream plane of R^+=550 open channel flow. (Wang and Richter 2019)

Figure 1.3: Flow Structures in turbulent boundary layers.

The characteristic features of VLSMs are relatively similar to LSMs, only being larger in size and length scale. They are known to carry over 20 δ in stream-wise direction through turbulent boundary layer flow. As well as, these features are also found in pipe and channel flow with wavelength of 10-20 R (Kim and Adrian 1999; Monty et al. 2007). Throughout a simulation study done by Lee and Sung 2013 on a pipe and boundary layer to directly compare the characterisation of VLSMs in both flows, it is indicated that VLSMs in pipe flow were generally much longer (up to 30 δ than in boundary layer).

They explained this to the entrainment occurring in the external flow, causing a more frequent break down stream-wise coherence, thence limiting the length of the VLSMs (Dennis 2015).

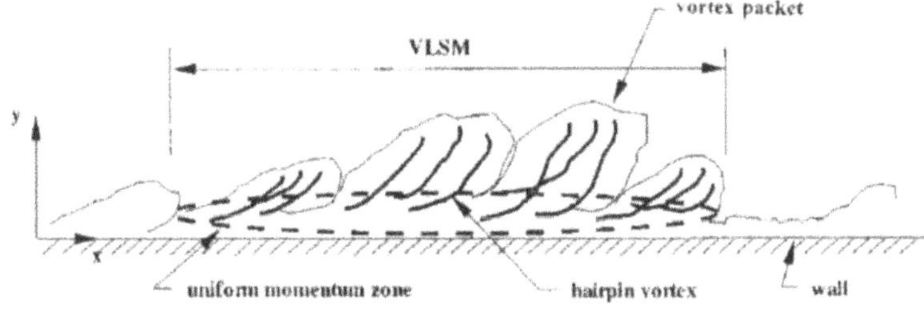

Figure 1.4: Conceptual model which describes the alignment of hairpins coherently into a package to form VLSM. (Kim and Adrian 1999)

Balakumar and Adrian 2007 provide the first evidence of VLSMs by using energy spectral analysis of pipe, channel, and boundary layers. Hutchins and Marusic 2007a demonstrated from the one-dimensional pre-multiplied energy spectra of the stream-wise velocity fluctuations that the contribution of the VLSMs to these fluctuations is Reynolds number dependence. Regions of high- momentum fluid have been observed in the logarithmic and wake regions of wall flows, and these regions are assigned to the VLSMs (Kim and Adrian 1999; Guala et al. 2006). Although the very large scale motions are found to be persisted well into the outer layer in the internal geometries (Bailey and Smits 2010), Further research on very large scale motions has discovered their influence on the near-wall small structures (Hutchins and Marusic 2007b; Abe et al. 2004).

As well, Mathis et al . 2009 and simulation results of Schlatter et al. 2009, mentioned evidence that the fluctuations of large scales in the log region are responsible for enhancing the amplitudes of small scales near the wall. This amplitude modulation was observed to increase with increasing Reynolds number over the Reynolds number range $Re_\tau \approx 10^3 - 10^6$. This relation considered LSMs and VLSMs as the most active and energetic structures that significantly contributed to turbulent kinetic energy and Reynolds stress production (Guala et al . 2006; Balakumar and Adrian 2007). However, they found that large structures greater than 3R contribute with 40-65 % of total kinetic energy and 30-50% of Reynolds shear stress. It is worth knowing that the Reynolds number plays an important role in detecting the region of energy production in-wall turbulent layers. This clarification was explained by Smits et al. 2011, who indicated from spectral analysis, low Reynolds numbers justified by the fact that the high kinetic energy occurs within the viscous buffer layer at a wall-normal distance of approximately y^+ $(yu_\tau/\nu) = 12$, while at high Reynolds numbers, the major contribution to the bulk turbulence production comes from the logarithmic region.

Furthermore, Bailey and Smits 2010 suggested that in the outer layer (beyond the logarithmic region), the hairpin packets comprise detached eddies, which have little correlation with flow near the wall, and this occurs across a wide range of azimuthal scales. On the other side, within the logarithmic region, it appears that hairpins are likely to be attached to the wall. Based on that, Bailey and Smits classified the LSMs into two classes; near-wall attached LSMs and outer-layer detached LSMs (beyond log region). Such classification recognised VLSMs to be only created due to detached LSMs aligning in the outer layer. This aligning is considered if the stream-wise alignment of LSMs causes the VLSMs. Whereas the attached LSMs appear to carry smaller transverse length-scale and smaller convective velocity in comparison with features of very large scale motions. Hence attached LSMs are unlikely to be involved in the formation of VLSMs.

Understanding the behaviour of coherent structures, including large and very large scale motions, particularly at relatively high Reynolds numbers, still attracts considerable attention due to its importance in practical applications.

1.1.3 Pipe Flow Boundary Layers

The fully developed turbulent flow in a pipe is proposed by Tennekes 1968; Afzal and Yajnik 1973; Afzal 1976 to be consisting of three main layers: outer, intermediate, and inner layers. In the terminology of classical theory, the inner layer includes the viscous sub-layer and the buffer layer. The intermediate layer encompasses the buffer region, logarithmic region, and the transition domain between the two regions. Furthermore, the outer layer contains the log region and wake region. It has been long recognised that the viscous sub-layer is at a distance from a wall $y^+ =$ y $u_\tau/\nu < 30$, where u_τ is the friction velocity, y is the wall-normal location, and ν is the kinematic viscosity.

The viscous sub-layers contain a buffer layer 3< y^+<30 and a linear sub-layer near the wall y^+<3. In the buffer layer, the Reynolds and viscous stress act directly on the mean flow, while in the linear sub-layer, the viscous stresses are dominant. Lately, the definition of turbulent flow layers has indicated an additional so-called overlap region. This region consists of an inertial sub-layer (y^+ >300) where the flow is nearly inviscid and a mesolayer (30<y^+<300) in which viscous stresses are neglectable.

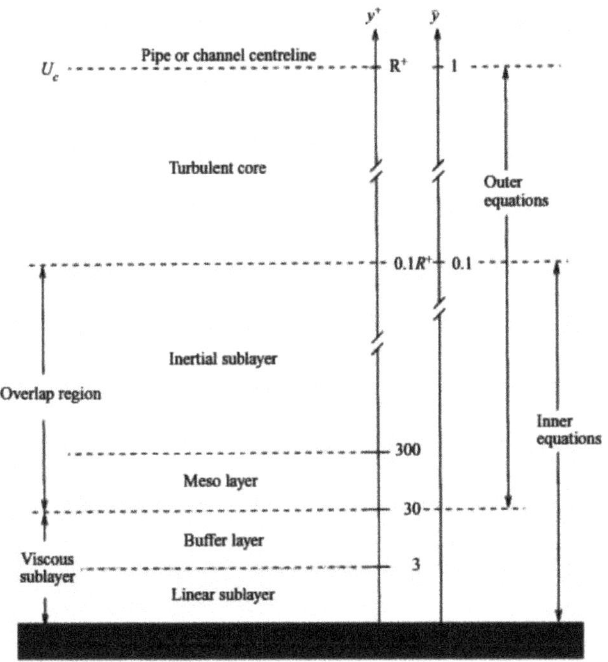

Figure 1.5: Schematic sketch of various regions and layers of pipe and channel flows, adopted from Wosnik et al. 2000.

Fig.(1.5) shows a schematic of various regions and layers of pipe flow. At very low turbulence Reynolds numbers, the near-wall region is considerably more interesting. However, the energy-containing eddies are constrained by the proximity of the wall. In this region, the dissipative and energy-containing ranges nearly overlap. The latter produces the Reynolds shear stress and feels the influence of viscosity directly. Here, the energy and dissipative scales are about the same. So, the dissipation ϵ is more reasonably estimated by $\epsilon \approx \nu q^2/\mathbf{j}^2$. Where, $\mathbf{j} \approx$ R/2 and q $\approx 3u_\tau$. Furthermore, for high Reynolds numbers $Re_\tau > 5000$, the energy dissipation is mostly determined by the large energetic scales of motions. These scales are effectively inviscid but control the energy transfer through the non-linear interactions (the energy cascades) to much smaller viscous scales where the actual dissipation occurs (Tennekes and Lumley 1972).

At the bottom of the overlap region, there will always be a mesolayer below about $y^+ \approx 300$, the dissipation and as well the Reynolds stress can never become independent of viscosity. The overlap layer is known as the averaged equations for the mean flow and Reynolds stresses from $y^+ > 30$ to the flow centreline, the averaged momentum equation is given approximately by :

$$0 = \frac{-1}{\rho}\frac{dp}{dx} + \frac{\partial(-uv)}{\partial y} \tag{1.1}$$

The Reynolds shear stress declines linearly to the center of the flow (Perry and Abell 1975), to become nearly constant outside y^+ =30. The pressure gradient vanishes in the constant Reynolds shear stress region at infinite Reynolds numbers. So the mean momentum equation reduces to :

$$0 = \frac{\partial(-uv)}{\partial y} \tag{1.2}$$

From both inner and outer momentum equations, the Reynolds shear stress gradient appears to be the common term in the overlap region that corresponds to this constant Reynolds shear stress layer (Wosnik et al. 2000). The Reynolds stress is effectively constant at finite large Reynolds numbers. At low Reynolds numbers, experiments do not have a region with constant Reynolds stress because the pressure gradient term is not truly neglectable. Although, it is observed that the viscous diffusion term is neglectable in the mean momentum equation, whereas viscosity does not appear directly in any of the single point equation governing the overlap region nor on those governing the outer layer.

1.2 Theoretical Background

To understand the experimental results of conical flows, i.e. turbulent channel and pipe flows, fundamental equations are needed to describe this phenomenon in a mathematical way.

1.2.1 Pipe Flow

Continuity equation:

$$\frac{\partial u}{\partial x}+\frac{\partial v}{\partial y}+\frac{\partial w}{\partial z}=0 \tag{1.3}$$

Momentum equation:

$$\rho\frac{d\overrightarrow{v}}{dt}=-\nabla p+\rho g+\mu\nabla^{2}\overrightarrow{v} \tag{1.4}$$

Mean of a turbulent velocity ($\bar{u}$)

$$\bar{u}=\frac{1}{T}\int_{0}^{T}udt \tag{1.5}$$

Fluctuating velocity ($\bar{u'}$), Turbulent intensity ($\bar{u}^{2+}$), and Time (T).

$$\bar{u'}=\frac{1}{T}\int_{0}^{T}u'dt=\frac{1}{T}\int_{0}^{T}(u-\bar{u})dt=0 \tag{1.6}$$

$$\bar{u}^{2+}=\frac{1}{T}\int_{0}^{T}u^{2+}dt\neq 0 \tag{1.7}$$

Since,

$$u=\bar{u}+u';v=\bar{v}+v';w=\bar{w}+w';p=\bar{p}+p' \tag{1.8}$$

By substituting in the continuity equation.

$$\frac{1}{T}\int_{0}^{T}\left(\frac{\partial\bar{u}}{\partial x}+\frac{\partial\bar{u'}}{\partial x}\right)dt+\frac{1}{T}\int_{0}^{T}\left(\frac{\partial\bar{v}}{\partial y}+\frac{\partial\bar{v'}}{\partial y}\right)dt+\frac{1}{T}\int_{0}^{T}\left(\frac{\partial\bar{w}}{\partial z}+\frac{\partial\bar{w'}}{\partial z}\right)dt=0 \tag{1.9}$$

Considering the similar analogy for other terms.

$$\frac{\partial \bar{u}}{\partial x} + \frac{\partial \bar{v}}{\partial y} + \frac{\partial \bar{w}}{\partial z} = 0 \tag{1.10}$$

The momentum equation in x-direction takes the following form;

$$\rho \frac{d\bar{u'}}{dt} = -\frac{\partial \bar{p}}{dx} + \rho g_x + \frac{\partial}{\partial x}\left(\mu \frac{\partial \bar{u}}{\partial x} - \rho \overline{u^{2+}}\right) + \frac{\partial}{\partial y}\left(\mu \frac{\partial \bar{u}}{\partial y} - \rho \overline{u'v'}\right) + \frac{\partial}{\partial z}\left(\mu \frac{\partial \bar{u}}{\partial z} - \rho \overline{u'w'}\right) \tag{1.11}$$

$-\rho \overline{u'u'}, -\rho \overline{u'v'} and - \rho \overline{u'w'}$ are called turbulent stresses.

by neglecting other components of turbulent stresses (not dominant) from Momentum equation

$$\rho \frac{\bar{u}}{dt} \approx -\frac{\partial \bar{p}}{\partial x} + \rho g_x + \frac{\partial \tau}{\partial y} \tag{1.12}$$

$$\tau = \mu \frac{\partial \bar{u}}{\partial y} - \rho \overline{u'v'} = \tau_{lam} + \tau_{tur} \tag{1.13}$$

1.3 Objectives of the Thesis

The dissertation aims to provide a better understanding of large and very large scale motions in turbulent pipe flows. This aim will be explored for LSMs and VLSMs in terms of its length-scale and energy contents and to define their contribution to total kinetic energy and Reynolds shear stress. In order to examine these goals in the 28 m long CoLaPipe test facility, several measurement campaigns are acquired as follow:

- High spatial measurements performed by single hot-wire anemometry at relatively high frequency to define the length scale of the LSMs and VLSMs and energy contents of the large and very large scale structures in fully developed turbulent pipe flow. The HWA datasets are analysed via the four principle statistical moments, spectral curves, contours, and cumulative fraction curves.

- High-speed PIV system is used to visualise and investigate large structures at low and high momentum regions in turbulent flow at the two-dimensional plane (x,y) through long enough captured (1838) snapshots of turbulent flow structures.

- Spectra of stream-wise velocity fluctuation, cospectral, and POD analysis for the PIV two-dimensional database are utilised to determine the energy content of scale structures of u, v velocity components.

1.4 Outline of the Thesis

Chapter 2 describes the CoLaPipe test facility in Brandenburg University of Technology Cottbus- Senftenberg used during the investigation. In addition, measurement techniques and their calibration methods were performed during the investigation, including Hot-wire anemometry (HWA) and High-Speed particle image velocimetry (HS-PIV).
Chapter 3 presents the two recognised calibration methods for (CTA) probe valid at low and high speeds covering a range 5 m/s $\leq u \leq$ 60 m/s to investigate the influence of both calibration methods on the consequences of hot-wire measurements. In accordance with each calibration method, hot-wire measurements are carried out in the CoLaPipe facility utilising the same single straight general purpose hot-wire probe (Dantec 55P11). Hot-wire data measurements of both calibration methods are compared throughout logarithmic velocity profile and statistical moments to identify that Ex-situ calibration induces a maximum error of 2.5-3.5 $\pm$ 0.5% comparing with In-situ calibration.

Furthermore, in chapter 4, one-dimensional spectral analysis is used to assess the structures behaviour in the outer region of pipe flow. The results of the power and pre-multiplied spectrum of stream-wise velocity indicate that the wavelength value of very large scale motions (VLSMs) acquires 19R at a maximum Reynolds number range $Re_D = 1 \times 10^6 (Re_\tau{=}19000)$. On the other hand, large-scale motions (LSMs) have a wavelength value of 3R over different Reynolds number range. Regarding the identified wavelength values, it is observed that contribution to energy for structures greater than 3R carries 55 % of total kinetic energy.

Measurements of time-resolved high-speed particle image velocimetry (HS-PIV) are applied at moderate and high Reynolds numbers in CoLaPipe to capture high-spatial instantaneous snapshots of recently explored large coherent structures in log and outer regions of fully developed turbulent pipe flow. PIV measurements in a stream-wise wall-normal plane are validated by comparing it with previous single hot-wire anemometry datasets that reveal a good agreement. Among visual inspection, stream-wise coherent structure features are exhibited in the middle of the log region and at the near-wall region as explained in chapter 5.

In chapter 6, time-resolved Particle Image Velocimetry (PIV) measurements are performed in CoLaPipe to investigate the contribution of large and very large scale motions to turbulence statistics at relatively high Reynolds numbers. Spectra statistical analysis is utilised to estimate the contribution magnitude of stream-wise/ wall-normal velocity fluctuations to total kinetic energy and Reynolds shear stress in the logarithmic and outer layer.

Finally, in chapter 7, a plane of snapshot proper orthogonal decomposition (POD) analysis of the turbulent pipe flow at Reynolds numbers, Re_τ= 6650 and Re_τ= 10617 are presented. The two-dimensional particle image velocimetry data are acquired in a stream-wise plane for a flow field 1-2 pipe radii. Energy contributions in the POD modes are explored to understand the spatial-temporal characteristics of the coherent structures.

Chapter 2: Experimental Facilities and Measurements Techniques

Turbulent flow investigations in pipes have been studied in earlier decades by Osborne Reynolds, who was a pioneer in studying fluid mechanics engineering. In 1877, he described in his published paper (Paper 24 in Reynolds 1900) methods for visualising fluid motions using filament dye. He applied this method to study vortex motion, considering vortex lines and vortex rings at several velocities. This technique for visualisation was utilised to determine the transition state of flow from laminar to turbulent flow in pipes. The famous experimental setup fig.(2.1) of Osborne Reynolds, as shown from Reynolds 1883, is still existing at the University of Manchester till today (Rott 1990), shows an innovative design of the original device. The figure indicates a siphon placed on an elevated platform at a distance high enough to increase the flow velocity and monitor the natural transition of the flow state. The exit valve is controlled by a lever that connected to the investigating pipe. The fluid (water) falls, together with a filament of dye, from a glass reservoir box into the investigating pipe. The tube has a trumpet-shaped inlet. This experiment was repeated 25 years later by V. Walfrid Ekman, who visited Manchester university to reuse Reynolds' original equipment (Rott 1990). After that, coming several researchers follow his methodology in different manners to complement the flow investigation. Table (2.1) shows the principle parameters of different test facilities with circular cross-sections, including their size, inner diameters, and Reynolds number range (Kármán number). As well as the table presents the length to diameter ratio L/D that explores if the flow is fully developed turbulent flow or not. However, the minimum developing length should be not less than 70 D proposed earlier by Patel and Head 1969 and later by König 2015 and Örgüner 2018 in CoLaPipe.

The introduced Reynolds number range for each facility, in fig.(2.2), is based on the friction velocity u_τ that is defined as $R_\tau = u_\tau$ R/ν, where u_τ is the wall friction velocity, R is the pipe radius, and ν is the kinematic viscosity. The selected various

Figure 2.1: Reynolds famous experimental setup (adopted from Rott 1990).

pipe facilities in table (2.1) represent the wall turbulence length scale $l_c = \nu/u_\tau$ as a function of Kármán number. This length scale is depending on the working fluid for each facility and its friction velocity. Thus, the spatial resolution differs from one facility to another. The widest Reynolds number range $Re_\tau = 1 \times 10^5$ is achieved by Princeton Superpipe with a limited spatial resolution $(0.3 \leq l_c \leq 25)$. This spatial resolution is different from CICLoPE (Centre for International Cooperation in Long Pipe Experiments). Although, it renowned with its large inner diameter D = 0.9 the viscous length scale of CICLoPE is about $(10 \leq l_c \leq 100)$ at a maximum Reynolds number range $Re_\tau = 40 \times 10^3$. Nevertheless, in CoLaPipe facility the two lines (suction and return line) allow measurements with high spatial resolution $(100 \leq l_c \leq 300)$ for its Reynolds number range $500 \leq Re_\tau \leq 20000$ (König et al. 2014).

Table 2.1: Various pipe test facilities.

Test facility	CICLoPE	CoLaPipe	Hi-Reff	Superpipe
Location	Bologna	Cottbus	Tsukuba	Princeton
Length (m)	111.5	28	9	26
Diameter (m)	0.9	0.19	0.1	0.13
max Re_τ	4×10^4	2×10^4	1.4×10^4	1×10^5
l_c @ max Re_τ	90	19	10	12
medium	air	air	water	air
Reference	Örlü et al. 2012	König 2015	Furuich et al. 2015	Marusic et al. 2013

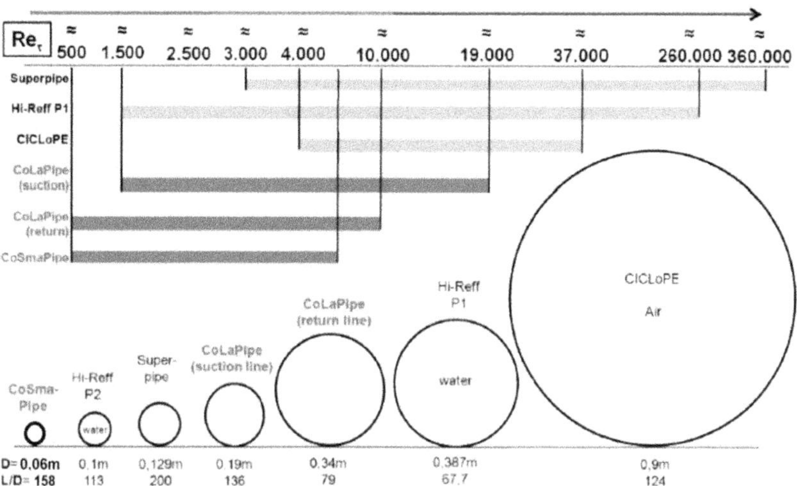

Figure 2.2: Comparsion between different pipe test facilites regarding aspect ratio (L/D), internal diameter and maximum Reynolds number range. (Öngüner 2018)

2.1 CoLaPipe

The current experimental measurements are represented in the Cottbus Large Pipe test facility (CoLaPipe) as shown in figures (2.3) and (2.4). The facility designed to investigate fully developed pipe flow at high Reynolds number range $6\times10^4 \leq Re_D \leq 1\times10^6$. The pipe produces flow whose mean properties are consistent within normal standards of experimental comparison with the results of numerous other pipe flow experiments. CoLaPipe is a closed-loop consists of two test sections representing in suction pipe section with inner diameter D = 190 ± 0.23 mm and return line with D = 340 ± 0.32mm. The two available test sections provide a length to diameter ratio of L/D = 148 and L/D = 79, respectively.

The main components of CoLaPipe are performed in the settling chamber, the inlet contraction, the blower, bends, and diffusers. The settling chamber is designed to eliminate the flow disturbances and ensure that the inlet flow velocity profile is in a uniform condition before entering the pipe test section with low turbulence intensity, less than 1%. Inside the settling chamber, passive flow control devices are consisting of four parts: a performed plate with rectangular passage 10 ×10 mm^2, a structured honeycomb plate with diameter d = 6mm and a total length (l) of 80mm to extend a length to diameter ratio of l/d = 12, a third essential flow controlling device represent in the five anti-turbulence screens which have a solidity of 40% and fixed to a metallic frame (König 2015). The minimum separation distance between two consecutive screens is 20 times the mesh size of the honeycomb. The last part of the control devices in the settling chamber that separates between the screen assembly and the inlet contraction is the relaxation part. However, this part works to provide a constant flow speed over the cross-section. The contraction of CoLaPipe with a high area ratio C_R =9 is joined between the settling chamber and the pipe test section. The contraction is made of glass fibre and designed to accelerate the flow coming from the settling chamber and reduce the variations of the mean axial velocity in the symmetrical plane. At the end of the suction pipe section, the power assembly is connected. The power assembly provides a flow rate of 0.05-2.5 m^3/s with a maximum velocity of 80 m/s at the contraction exit.

In order to stabilise the working temperature inside both test sections, the CoLaPipe facility is equipped with a heat exchanger to keep the temperature controlled within a range of 15 – 21°C and a maximum deviation of ΔT = ± 0.5 K at the actual measuring position. For more details about CoLaPipe components, see König et al. 2014 and König 2015. All the segments of the two pipe sections are made of transparent plexiglass to facilitate optical visualisation. In addition, one or two glass pipe segments can be used as a test section in the suction line for LDA and PIV measurements to acquire data with sufficiently high resolution. The set-up of the HS-PIV measurements is described minutely in chapter (5). All experimental measurements are carried out at location L/D = 110 in the suction pipe section. In order to ensure that the flow is fully developed, adequate tripping devices such as orifices are settled at exit contraction.

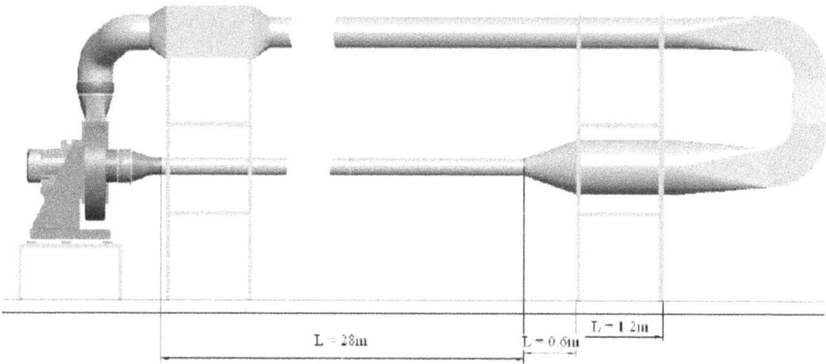

Figure 2.3: Sketch of the CoLaPipe at the Department of Aerodynamics and Fluid Mechanics (LAS BTU Cottbus-Senftenberg). (König 2015)

Figure 2.4: The 28 m long CoLaPipe facility in LAS.

According to König 2015, the development length is estimated for the natural transition within the CoLaPipe to be x/D = 70 . Fig.(2.5) shows the development of statistical quantities along the pipe test section for two different Reynolds numbers. Therefore, x/D $\geqslant$ 70 is considered to be fully turbulent. König 2015 finding agrees with Patel and Head 1969, who verified a minimum development length in pipe flow.

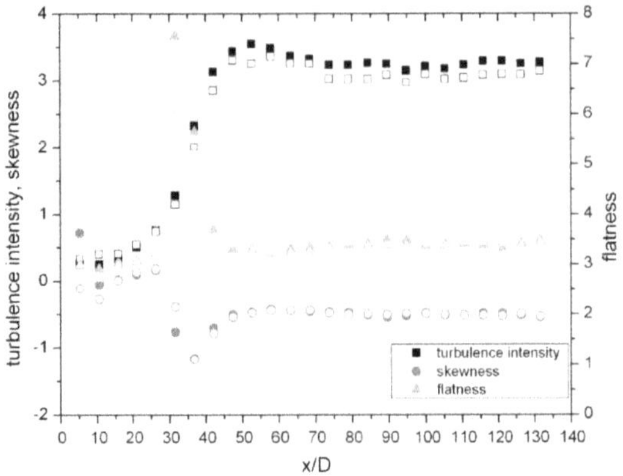

Figure 2.5: Development of statistical quantities along pipe test section at $Re_D = 1.8 \times 10^5$ (open symbols) and $Re_D = 4 \times 10^5$ (filled symbols). (König 2015)

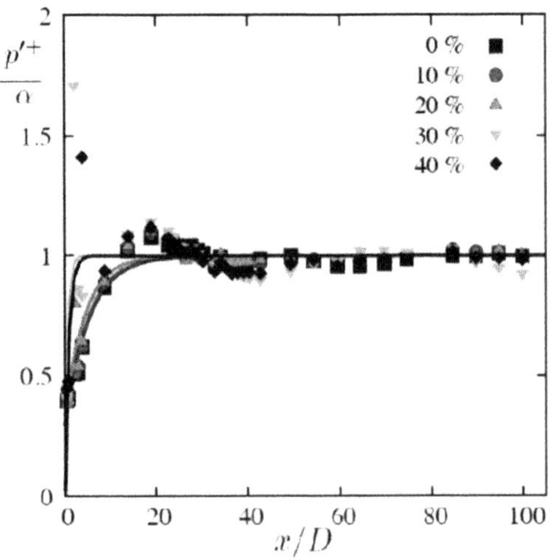

Figure 2.6: Wall pressure fluctuations at $Re_D \approx 300.000$ along the CoLapPipe for different ring disturbances (Selvam et al. 2018)

On the other side, Selvam et al. 2018 and Öngüner 2018 used the ring disturbance (orifice) in CoLaPipe with different blockage ratio for reaching the earlier flow transition artificially, i.e., 0% disturbance, the position of saturation is x/D $\approx$ 50, whereas in case of 20% disturbance (red line), the position of saturation shifts upstream x/D $\approx$ 34 as in fig.(2.6).

2.2 Measurement Techniques

2.2.1 Hot-Wire Anemometer

Hot-wire anemometry (HWA) is a standard and well-established technique for measuring the velocity of airflow. The main purpose of the operation is based on the heat transfer between the micrometre size wire heated by an electric current and the airflow passing over it. The cooling effect from the airflow changes the temperature of the wire and its resistance. For keeping the temperature constant, electric energy is required. These quantities can be measured through a bridge circuit. Following calibration, the relation between the voltage and the flow speed can be derived. The calibration has to be carried out in conditions similar to that of the experiment (e.g. Ex-situ calibration), and sometimes In-situ calibration is recommended. The influence of Ex-situ and In-situ calibration methods on hot-wire measurements is clarified well in chapter (4). Keeping the electronic noise in the electronic circuit as low as possible, whereas The feedback loop in the electronic circuit of hot-wire is adjusted for an optimum frequency response for a particular flow. Hot-wires are usually made of 2.5 or 5 μm tungsten wire.

For incompressible flows at Mach number less than 0.3 as in CoLaPipe facility condition, by definition, the density is constant, and therefore the hot-wire captures the velocity directly. The hot-wire is simply made of a very thin wire welded to the tip of two support prongs. The sensor wire is usually of a diameter of 5 μm and a length of 1 mm. Some cases are used a shorter probe for better spatial resolution. To keep the length to diameter ratio (important parameter) high enough, the diameter is usually reduced to 2.5 μm. This size of wire is also more commonly used for high-speed flows. The obvious advantage of hot-wires represents in their high-frequency response, which makes them ideal for measuring unsteady flows at very high frequencies of the order of several hundred kilohertz.

2.2.2 Types of Hot-Wire Probes

Two major types of hot-wire probes are used in current experiments, which are suitable for measuring internal flow inside CoLaPipe.

- The single straight hot-wire probe

  The single straight hot-wire probe (55P11) from Dantec has straight prongs. Its sensor is perpendicular to the probe axis as shown in fig.(2.7.a). It is utilised to measure the mean and fluctuating velocities of one-dimensional velocity component. The hot-wire mounts so that the probe axis is parallel to the direction of the flow. The probe has a dimeter (ϕ) 5 micrometre and 1250 micrometre in length (l_w). The viscous scaled length is ranges $85 \leq l^+ \leq 247$.

- The single boundary layer hot-wire probe

  The boundary layer hot-wire probe (55P15) prongs from Dantec differ from the straight one. However, the prongs in the boundary layer have a right angle position, and the sensor is perpendicular to the probe axis with 90° illustrated in fig.(2.7.b). Such probe is suitable for boundary layer measurements, and it is used to measure the mean flow velocities and flow fluctuations in locations that are not easily reached as in pipes. This probe is mounting with the probe axis to be perpendicular to the direction of the flow.

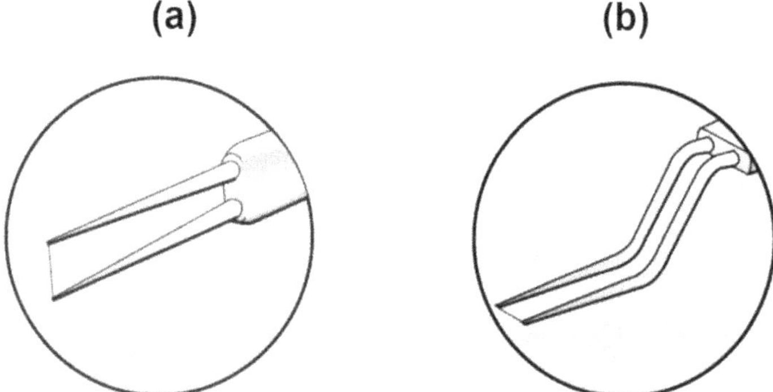

Figure 2.7: Hot-wire anometery probes.(a) Single straight hot-wire probe (55P11), and (b) Boundary layer hot-wire probe (55P15). Dantec website.

Hot-wire anemometer calibration is an essential preparation process that should be undertaken before carrying out any measurements. Overheat ratio adjustment, total resistance measurement, and ambient temperature are principle parameters required to acquire accurate acquisition data in high spatial resolution.

2.2.3 Overheat Ratio

The overheat ratio determines the sensor operating temperature. The overheat resistor, fig.(2.8), located at the right bridge arm is used to adjust the needed working temperature that can be reached when the bridge is operated. The cold and warm resistors are related via the overheating ratio, a_w:

$$a_w = (R_w - R_o)/R_o \tag{2.1}$$

where,
R_w: is the sensor resistance at operating temperature T_w.
R_o: is its resistance at ambient (reference) temperature T_{amb}.

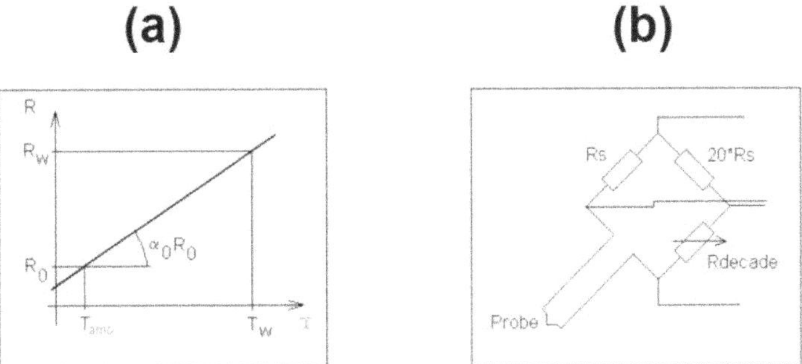

Figure 2.8: (a) Relation between sensor resistance and temperature, and (b) Wheatstone bridge circuit. Adopted from Dantec website.

The total resistance R_{tot} is measured at ambient temperature T_{amb} and calculate the active sensor resistance:

$$R_o = R_{tot} - (R_{leads} + R_{support} + R_{cable}) \tag{2.2}$$

where,

R_{leads} = probe leads the resistance,

$R_{support}$ = support resistance

R_{cable} = cable resistance.

The overheat recommended ratio for HWA probes in the air is 0.8, giving an over-temperature between 200 and 300°C.(Dantec website)

2.2.4 Hot-Wire Anemometer Velocity Calibration

Calibration of hot-wire anemometer probe is used to establish a relation between the output voltage of the probe (E) and the velocity of the flow (u). It is performed by exposing the probe to a stream flow with known velocities to measure the output voltages. A curve is used to fit the relation through the measured points (E, u) represents the function to be used when converting data records from voltages to velocities.

Calibration can either be carried out in a dedicated external probe calibrator (Ex-situ), which normally is a free jet or inside the test section of a facility using a pitot-static tube as a reference for the velocity (In-situ). It is recommended to record of the temperature during calibration to consider the temperature variations during calibration.

2.2.5 Polynomial Curve Fitting

After Plotting u as a function of E_{corr}, a polynomial trend line is created in fourth order:

$$u = C_0 + C_1 E_{corr} + C_2 E_{corr2} + C_3 E_{corr3} + C_4 E_{corr4}, \tag{2.3}$$

where C_0 to C_4 are the calibration constants.

The importance of polynomial curve fit is that it makes very good fits with linearisation errors of less than 1%.

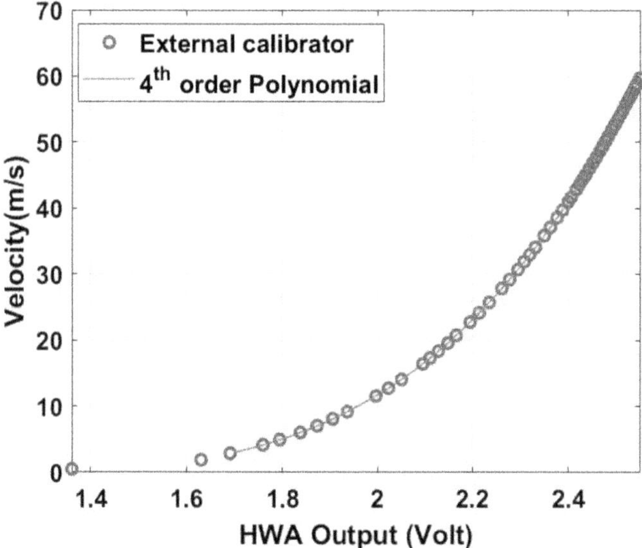

Figure 2.9: Hot-wire anemometry calibration curve.

2.2.6 Power Law Curve Fitting

Another method that can be used for curve fitting is called Power law curve fitting by using the King's Law. In this method, the measured volt E(u) is plotted. The calibration constants A and B are deduced by linear regression of the measured data, considering the value of n to be 0.45.

$$u = (E/B - A)^{1/n} \tag{2.4}$$

Changing the value of n and repeat the calculation till the deviation between the measured and the calculated data is minimized. The accuracy of the power law method is the less accurate compared to the fourth order polynomial fitting method, especially over wide ranges of Reynolds numbers, whereas the estimation of n shows Reynolds number dependent.

2.2.7 Particle Image Velocimetry (PIV)

Particle Image Velocimetry (PIV) is one of the none intrusive modern measurement techniques used to measure the fluid flow by sensing the displacement of artificially injected particles convected by the flow in high spatial and temporal resolution. The PIV was known as Laser Speckle Velocimetry (LSV), and many references were published on its development since 1977, particularly applying it to turbulent flow (Dudderar and Simpkings 1977, Meynart 1983 a, and Meynart 1983 b). Later studies with a new technique, which at the time called Particle Image displacement velocimetry (PIDV), has been carried out (Adrian and Yao 1983, Lourenco 1984). Another progress carried out in digital imaging facilitated the technique of rapid evolution. Gauthier and Riethmuller 1988, successfully measure the three velocity components. In 1986, the first experiment using a video camera was presenting (Kompenhans and Reichmuth 1986) and became one of the most common popular techniques used in fluid mechanics research worldwide.

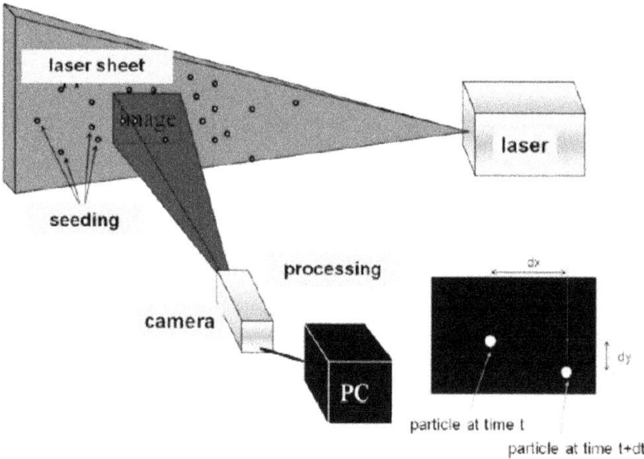

Figure 2.10: Schematic drawing of the PIV set up. Adopted from Chanetz et al. 2020.

The PIV setup, as shown in fig.(2.10), consists of a laser sheet (about 1 mm thick) expanded by a cylindrical rod that illuminates the measurement plane. The laser is pulsed continuously at a very short time delay to light up the particles seeded into the flow. The camera is positioned normal to the laser sheet plane and records the particles' displacement at two-time instants. The snapshots are then processed to determine the displacement of the particles within the 2D flow field to measure two velocity components in the laser sheet plane. The PIV system used in the present investigation is a Time-resolved high-speed PIV. This type of PIV allows to study the temporal evolution of flow and to explore the flow mechanisms in time and space. It can also help to provide information about instability modes and turbulent structures. For a time-resolved PIV, a Complementary Metal-Oxide Semiconductor (CMOS) camera with a high frame acquisition rate is required and a more powerful laser pulsing at a speed of similar order. The image pairs are correlated to extract the velocity and obtain a time series of the flow field where one can deduce the dynamic behaviour of turbulent flow in frequency domain or spectra.

2.2.8 PIV Main Components

- Nd: YLF laser:

  The ND: YLF lasers are used for several applications, including the high-speed PIV techniques, which require an efficient power laser source that allows convergent of the frequency to apparent wavelengths. The main element of the Nd: YLF laser is the neodymium: yttrium lithium fluoride (Nd^{3+}:YLF) crystals. These crystals can emit the highest pulse energy and average power, with frequency rates ranging up to 10 kHz. (Raffel et al. 2007)

2.2.9 How Nd: YLF Laser Emits the Laser Ray ?

The laser diodes supply energy to the active medium (neodymium-doped yttrium lithium fluoride crystals) to convert electrical energy into light. The optical resonator consists of two silvered mirrors, which are mounted in front and rear to the Nd: YLF crystal. The front mirror (fully silvered) works to reflect all the light, while the other mirror (partially silvered) reflects most of the light and allows a small portion of light through it to emit the laser beam as shown in fig.(2.11).

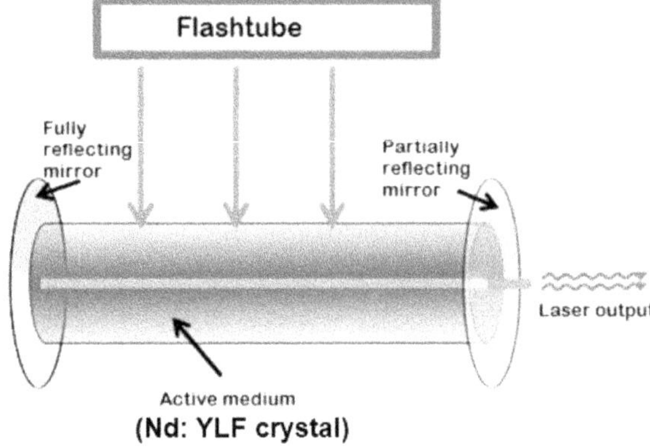

Figure 2.11: Schematic drawing for the Nd:YLF crystal. Raffel et al. 2007

- CMOS Camera:

  At the beginning of using the PIV, the images are recorded to photographic films which are less sensitive to the wavelength of the light source. These images were analysed directly or digitised using a scanner. Early 90's, sensors in the PIV acquisition system introduced to transfer the light collected (i.e. photons) into electric charges (i.e. electrons). The most frequent used sensors in modern PIV systems are Complementary Metal-Oxide Semiconductor (CMOS). The charges are converted into a voltage in the pixel. This exchange leads to recording the images carried out directly, pixel by pixel, over a very short period.

- Seeding particles:

  The particles seeded to the flow must be in the submicron range in order to be able to track flow gradients. One of the most important parameters is that seeding particles must be smaller as velocities increases. If the seeding particle size is too large, it will not be carried by the flow and cause a false representation of the flow. In case the particle size is tiny, the particle will not reflect enough light that will increase of uncertainty in the signal processing. Therefore, choosing a proper particle size is essential to get a precise measuring data.

  For generating particles, different solutions are possible: atomisers of various types, vaporisers, injectors, fluidised beds (Meyers 1991). These generators must comply with some requirements, such as enough particles must be produced with limited diameter. Also, the distribution size has to be stable with time.

2.2.10 Pressure Gradient Measurements

For measuring steady and variance static pressure in the pipe flow, small orifices (taps) are mounted on the pipe wall surface to face the flow, then connecting them to a manometer or pressure transducer for data recording. Although this method is used widely, it may introduce appreciable systematic errors if the taps are not implemented correctly. One of these errors represents in the formation of counter-rotating vortices over a pressure tap.

Figure 2.12: Photograph of pipe test section provided with pressure tapping for measurement of static pressure along the pipe.

The hole size range of static pressure taps is usually between 0.5 and 3 mm to overcome the existence of these vortices.

2.2.11 Application at CoLaPipe Test Facility

- **Static pressure**

  Measuring pressure gradient is essential to identify the wall friction velocity u_τ. The friction velocity is defined as $u_\tau = (\tau_w/\rho)^{0.5}$, where τ_w is the wall shear stress which estimated as τ_w = -(D/4)(dp/dx) and ρ is the density of the working fluid. The CoLaPipe test facility is equipped with 14 static pressure measuring station to measure the pressure gradient (dp/dx), i.e. the distance between each pressure measuring station is one meter. The pipe is equipped with three pressure taps orientated around the circumference of the pipe segment at each station. However, the suction pipe section of 28 m long is provided with 14 pipe segments of 2 m length. The pressure taps are connected to a pressure scanner (PSI 9116 Ethernet Pressure Scanner) to record the static pressure at different inlet flow speeds. The stream-wise pressure gradient dp/dx is obtained by fitting the linear behaviour to the static pressure measurements in the range $5.2 < L/D < 142$.

- **Total Pressure**

  The stagnation or total pressure of a flow can be measured through a Pitot probe which is basically a tube with a small hole facing the flow. The pitot tube is connected to a pressure transducer to record stagnation pressure. The L-shaped Prandtl tube (pitot tube) is proposed to measure the centreline speed u_c of the pipe flow by placing the tube upstream at the contraction exit facing the flow at the pipe axis. On the other side, the bulk velocity u_b is measured via the pressure drop across pipe contraction and is calculated using the Bernoulli equation.

2.2.12 Fluid Parameters

- **The density of the air**

  The density of the air (ρ) can be calculated using the Perfect gas law:

$$\rho = P/\mathcal{R}T_{amb} \tag{2.5}$$

where,

P : is the total pressure in pascal (pa),
$\mathcal{R}$: is the gas constant.
T_{amb} : is the ambient temperature in Kelvin (K).

- **Kinematic viscosity**

  With the help of Sutherland's law, Kinematic viscosity can be estimated

$$\nu = \frac{C_1 \mathbb{T}^{3/2}}{\rho(\mathbb{T} + S_1)} \tag{2.6}$$

where, C_1 and S_1 are constants (1.458×10¯6 and 110.4 for air, respectively),
ρ : air density (kg/m^3),
$\mathbb{T}$ is the temperature of the fluid in Kelvin (K).

Chapter 3: Influence of Calibration Methods on HWA Measurements

3.1 Theoretical Background

The experimental investigation of turbulent flows has been performed previously in several studies to a large extent employing hot-wire anemometry (HWA) which is an unrivalled technique compared to others due to its high-frequency response and temporal resolution. However, this technique has several limitations, particularly in wall-bounded turbulent flows or complex flow, such as wall interference and heat conduction to the wall. It is still expected to continue to be used for the coming decades to study fluid dynamics and constitute a reference to other experimental measuring techniques due to its accurate interpretation of the signal and simplicity in use. Hot-wire Anemometer measuring technique is an indirect method. This method needs to be calibrated against a known velocity that is related to the voltage reading from the HWA system. The accuracy of hot-wire measurements is dependent on the accuracy of the calibration that relates to measured heat transfer from the hot-wire to the velocity through the calibration curve (Talluru et al. 2014). Two recognised methods (In-situ and Ex-situ) of calibrating constant temperature hot-wire anemometers (CTA) for velocity measurements have been proposed and used by researchers. Durst et al. 2008 explained that HWA is sensitive to the instantaneous volume flow rate and to the gas heat conductivity. Consequently, any errors in the calibration process increase uncertainty in the hot-wire measured velocity measurement (Talluru et al. 2014). In the present study, the two calibration methods are conducted to explain whether the calibration method plays an important role in the hot-wire measurement results accuracy or not? This accuracy in measurements will be revealed by comparing the results of both methods in terms of the statistical moments and the logarithmic velocity profile of hot-wire measurements. The experiment is carried out at two relatively high Reynolds numbers $Re_\tau = 6600$ and $Re_\tau = 13000$ for each calibration method, where Re_τ is the Reynolds number based on the friction velocity (u_τ) and the pipe radii (R).

3.2 Experimental Setup

The experimental data is obtained in the CoLaPipe test facility located at Aerodynamics and Fluid Mechanics department shown in fig.(3.1), The pipe flow facility is 27 m long Plexiglas pipe with diameter D = 190 mm. On account of calibration drifts that may be caused due to changes in the ambient temperature of the flow (Comte-Bellot 1976), the CoLaPipe facility is equipped with a heat exchanger (temperature controlled with ± 0.5 °C. Air passes through a settling chamber, a honeycomb, and a grid from a blower before it enters the pipe.

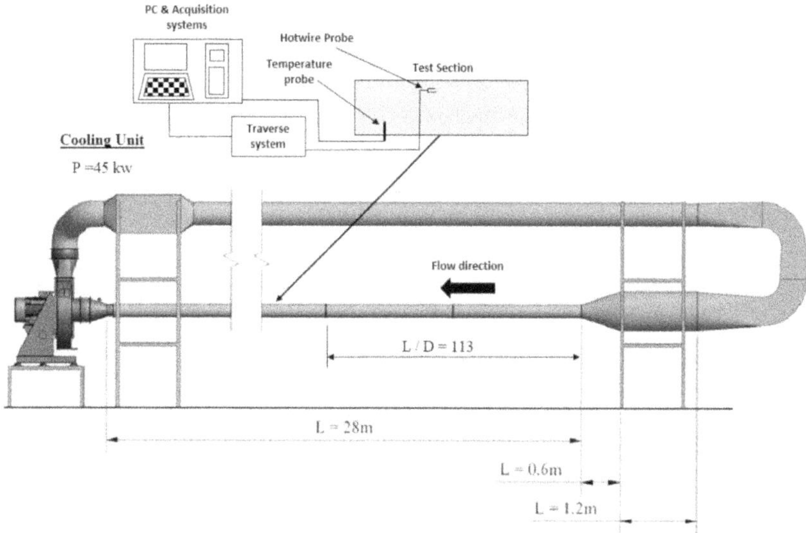

Figure 3.1: Sketch of the HWA experimental setup in CoLaPipe test facility. (König 2015)

Furthermore, calibration drifts could be caused as a result of changes in other ambient conditions such as humidity (Durst et al. 1996), pressure; wire degradation due to incomplete cleaning (Willmarth et al. 1977), electro-migration (Willmarth et al. 1984);

dust particles in the flow (Collis 1952) and wire fouling (Wyatt 1953). Therefore, several correction schemes are available, if it is needed for calibrations (e.g. (Bruun 1995, Cimbala and Park 1990, Durst et al. 1996, Hultmark and Smits 2010, Nekrasov and Savostenko 1991, Perry 1982, Tropea et al. 2007).

A standard single straight hot-wire probe (Dantec 55P11) is utilized for all measurements with a diameter (d_w) 5 μm, length (l_w) 1250 μm provides length to diameter ratio (l_w/d_w) 250, viscous scaled wire length range $85 \leq l^+ \leq 247$, and maximum temperature resistance (T_w) $\approx$ 260-270 °C. The hot-wires are operated in constant temperature anemometer (CTA) mode via a Dantec StreamLine Pro frame and 91C10 CTA channels with an overheat ratio $(a_w) = (R_w\text{-}R_o)/R_o = 0.8$, where R_w is the hot-wire resistance at operating temperature and R_o is the resistance at reference (ambient) temperature. The sampling frequency is set to 20 kHz. As mentioned previously, the hot-wire calibration is accomplished via two renowned methods. The main reason for naming the two calibration methods is based on the HWA measurement position and the surrounding flow field conditions. The probe calibration is performed In-situ by placing the hot-wire probe at the centreline of the pipe to face the flow, at its laminar state, just at the contraction exit.

The temperature of the pipe flow is approximately constant measured through a Dantec Compact temperature thermocouple. On the other side, the Ex-situ calibration is performed where the hot-wire probe is developed about one diameter (10 mm) away from the exit of the nozzle in the middle of the external calibration jet. This can be shown in fig.(3.2). The velocity of the nozzle flow is controlled by the flow controller, which is supplied with compressed air. This calibration is done at the same In-situ calibration velocity range. The jet velocity is evaluated from the pressure difference across the nozzle. The temperature of the jet flow is required for density evaluations, and it may also be used for temperature compensation. Therefore the temperature in the nozzle settling chamber is measured by the thermocouple connected to the jet unit.

For both calibration methods, a standard static calibration is carried out at the same velocity range of 5-60 m/s at a constant temperature, consisting of approximately 25 simultaneous values of hot-wire voltage E in volt spaced over the selected velocity range. During calibration, hot-wire voltage E, flow velocity u, and flow temperature $\mathbb{T}$ are acquired for each calibration point. The parameters are recorded via PC by using labVIEW

Figure 3.2: Left:Ex-situ calibration unit with air supply.Right:Calibration unit contraction with probe to be calibrated.

program for the two calibration methods, then all data are analysed using the Matlab program. The polynomial fits have the advantages that the calibrations are carried out in the format u= f(E) used for measurements. However, to realise the required accuracy, it is necessary to use either the fourth-order polynomial E or the third-order polynomial E^2 to generate calibration curves (Bruun et al. 1988). In the recent study, curves are fitted with a fourth-order polynomial, as suggested by Perry (1982). Lee and Budwig 1991 indicated that the velocity data obtained by calibration jet is about 4-50 % lower than the real velocity essentially at low-velocity values $<$ 2 m/s. Yue and Malmström 1998 proposed to use the laminar pipe flow method (In-situ method) for hot-wire measurements in the low-speed range of 0.1 m/s and above.

Furthermore, Johnstone et al. 2005 used the same method to calibrate a hot-wire in the velocity range of 0.125-13.78 m/s. Consequently, the recent investigation is taken In-situ calibration as a reference due to its accuracy in measuring turbulent flow at real speed values. During the calibration process, the hot-wire probe is oriented only to a laminar flow field. This equips the hot-wire to be more sensitive during measuring the flow throughout this calibration method. Fig.(3.3) presents the Ex-situ calibration curve compared to the In-situ one. Although the uncertainty measurements of the recalculated velocity $u_{recal.}$ is < 1 %. For both calibrations, a slight deviation of the Ex-situ curve is clearly observed. This phenomenon is frequently induced due to the surrounding flow field during the Ex-situ calibration procedure.

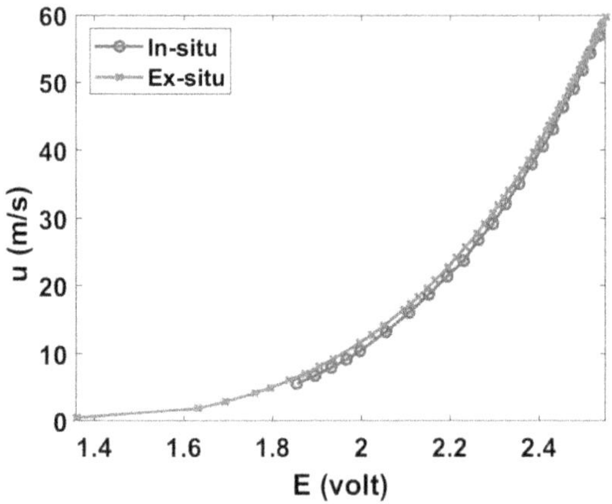

Figure 3.3: Calibration profiles for a single wire probe. o; In-situ calibration data points, x; Ex-situ calibration data points, –; 4th order polynomial.

After each calibration manner, hot-wire measurements are carried out in a test section located at L/D =110, where the flow is fully developed. The measurements are conducted in a radial direction for $Re_\tau = 6600$ and $Re_\tau = 13000$ at several wall-normal locations y/R (30 points). The recorded samples are 3.6×10^6 for each point. For hot-wire probe mobility, a computer-controlled traverse system was used to the HWA probe after the calibration process. The traverse is placed on a scaled rail to facilitate the movement in axial and radial directions. Same hot-wire anemometer probe is used to ensure the correctness of all measurements.

3.3 Results

The main purpose of the hot-wire measurements is to find out the influence of each calibration manner on the final results. Due to the high Reynolds number, a near-wall region can not be well resolved. Therefore, the recent HWA measurements are carried out in logarithmic layer and outer region. The hot-wire measurements which are based on the Ex-situ calibration method shows a distinguishable deviation at the two different Reynolds numbers. With increasing the Reynolds number $Re_\tau = 13000$, the deviation increases obviously towards the pipe axis at wall-normal locations $y^+ > 1300$, this clarified well in fig.(3.4).

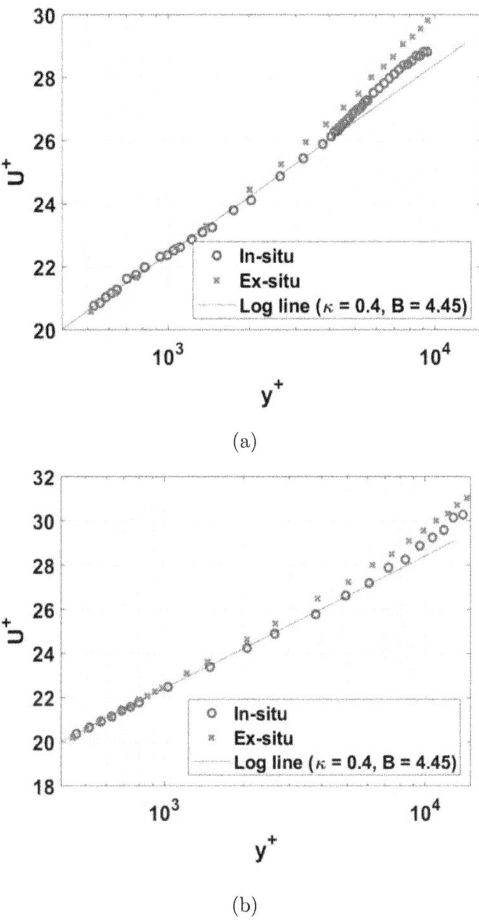

Figure 3.4: Logarithmic velocity profiles for In-situ and Ex-situ calibration methods. (a) $Re_\tau = 6600$, and (b) $Re_\tau = 13000$.

For evaluating the effect of HWA calibration on measurement accuracy, the present measurements are validated through the logarithmic velocity profile. The velocity profile behaviour of the two HWA measurements is roughly closed to each other, particularly at the wall-normal locations towards the wall. Fig.(3.4) clarifies that the present data collapse with the logarithmic line $u^+ = 1/k\ ln(y^+)+B$. Subsequently, it starts to separate near the pipe centreline deducing a salient deviation of 2.5±0.5% at $Re_\tau = 13000$. This deviation proceeds from the measurement data related to the Ex-situ calibration method. Moreover, the deviation also exists at the lower Reynolds number value but with higher amplitude of 3.5±0.5%. The usable von Kármán's (k) = 0.4 and additive constant (B)= 4.45 were proposed by Balley et al. 2014.

Furthermore, at two measured Reynolds numbers $Re_\tau = 6600$ and $Re_\tau = 13000$, The behaviour of root mean square profile (RMS) indicates a noticeable deviation via two-coloured curves in fig.(3.5). The blue and red profiles refer to comparable recent measuring data corresponding to In-situ and Ex-situ calibration methods, respectively. A significant deviation is emerged at $Re_\tau = 6600$ than the higher Reynolds number $Re_\tau = 13000$ to accomplish a deviation range of 10-15 ±0.5%.

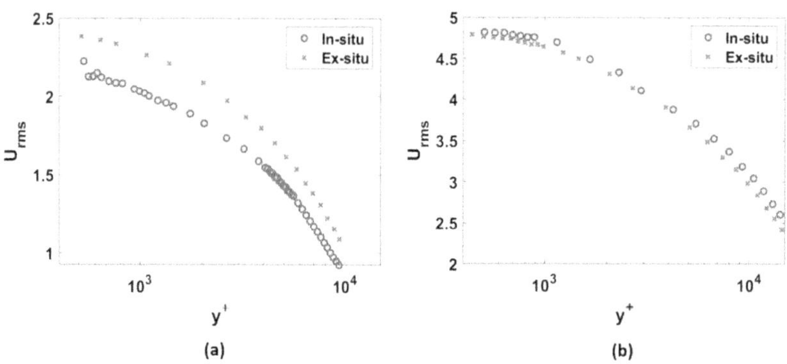

Figure 3.5: (a) Root mean square (RMS) profiles for In-situ and Ex-situ calibration methods at $Re_\tau = 6600$, and (b) RMS profiles for In-situ and Ex-situ calibration methods at $Re_\tau = 13000$.

The influence of different calibration methods can be presented through the turbulent intensity in fig.(3.6). From the figure, it is noticed that the inner peak cannot be detected due to the effect of high Reynolds number where the boundary layer thickness shrinks or collapses towards the wall region. Therefore, the measurement datasets represent solely the outer peak compared with the dataset of the Nano-Scale Thermal Anemometry Probes (NSTAP) (Zagarola and Smits 1998). Based on this comparison, it is found that by matching the In-situ hot-wire data of $Re_\tau = 6600$, a strong convergence is realised at $y^+ \geq 527$, while the Ex-situ of similar Reynolds number indicates a high deviation compared with the other given data.

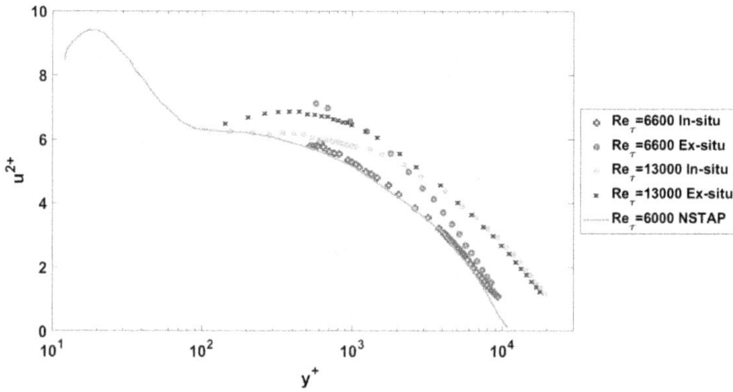

Figure 3.6: Turbulent Intensity velocity profile (u'^2/u_τ^2) for the In-situ and Ex-situ datasets at $Re_\tau = 6600$ and 13000 compared to the data from Zagarola and Smits 1998.

On the other side, the In-situ dataset for higher Reynolds number Re_τ= 13000 starts to form a rough convergence at a range $154 \leq y^+ \leq 300$; however, this is not proved with the Ex-situ dataset. The plateau of the outer peak in the logarithmic region is barley observed for Re_τ =13000 than at the lower Re_τ =6600. This phenomenon has been explained previously by Marusic and Kunkel 2003. According to the turbulent intensity outputs, the In-situ calibration method indicated roughly equivalent results to NSTAP databases. No significant errors are observed with a higher Reynolds number, but about 21% are referred to the lowest Reynolds number.

Otherwise, for statistical moments represented in kurtosis $\sum_i^N \frac{(u_i-u_m)^4}{N(\delta)^4}$; the measure of the amplitude distribution (flatness factor), and skewness $\sum_i^N \frac{(u_i-u_m)^3}{N(\delta)^3}$; the measure of the lack of statistical symmetry in the flow. The calculated kurtosis of the data in fig.(3.7) presents a good collapse with proposed DNS data at Re_τ= 1080 (Penga et al.

2018) at all wall-normal locations (y/R) for the two Reynolds numbers. Concerning the statistical moments definitions, u_i and u_m are the local velocity and mean velocity respectively, while, δ is the variance $(\sum_i^N \frac{(u_i-u_m)^2}{N-1})^{1/2}$ and N is the number of samples.

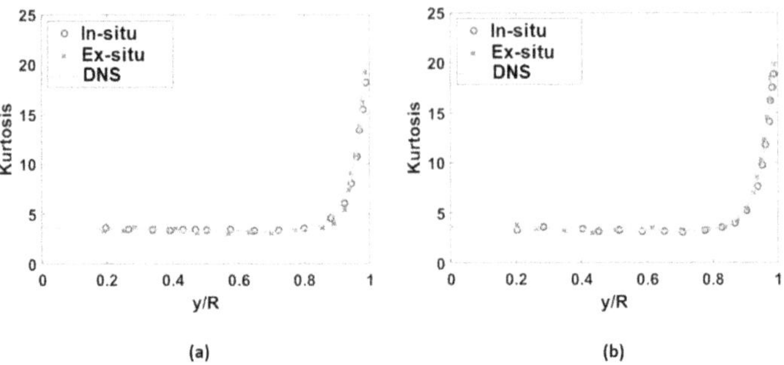

Figure 3.7: Kurtosis profiles for In-situ and Ex-situ calibration methods versus DNS data $Re_\tau = 1080$ (Penga et al. 2018). (a)$Re_\tau = 6600$, and (b) $Re_\tau = 13000$.

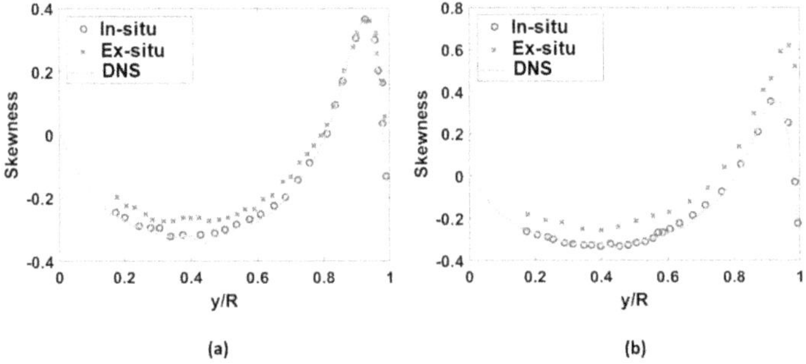

Figure 3.8: Skewness profiles for In-situ and Ex-situ calibration methods versus DNS data $Re_\tau = 1080$ (Penga et al. 2018). (a)$Re_\tau = 6600$, and (b) $Re_\tau = 13000$.

Skewness (S) data illustrates different consequences in fig.(3.8). The behaviour of the measuring data corresponding to In-situ performs good compatibility with the DNS data at $Re_\tau = 13000$, while a satisfactory agreement is accomplished at $Re_\tau = 6600$.

Undoubtedly, the calibration method plays an important role in resolving the spatial resolution that can be defined via single hot-wire measurements at different points. Based on the two calibration methods, the power spectra plots present in fig.(3.9) and (3.10) at selected wall-normal locations. The spectra are grouped into two separate plots to make this phenomenon clearer. Fig.(3.9.a and b) represents the power spectra of wall-normal locations close to the wall while further locations from the wall are indicated in fig.(3.9.c and d). For the lower Reynolds number Re_{τ}= 6600, slight variation in power spectrum between both methods is noticed at y/R= 0.12, 1, while in the logarithmic region, a significant deviation is shown at y/R=0.3, 0.76 as in fig.(3.9.b and c). This deviation could cause under/over estimating values of wavenumber at relatively moderate Reynolds number in intermediate and outer-layers. What is concluded from the present results based on using In-situ calibration data set as a reference which represents the most precise method for hot-wire calibration methods comparing with Ex-situ as deduced from fig.(3.6).

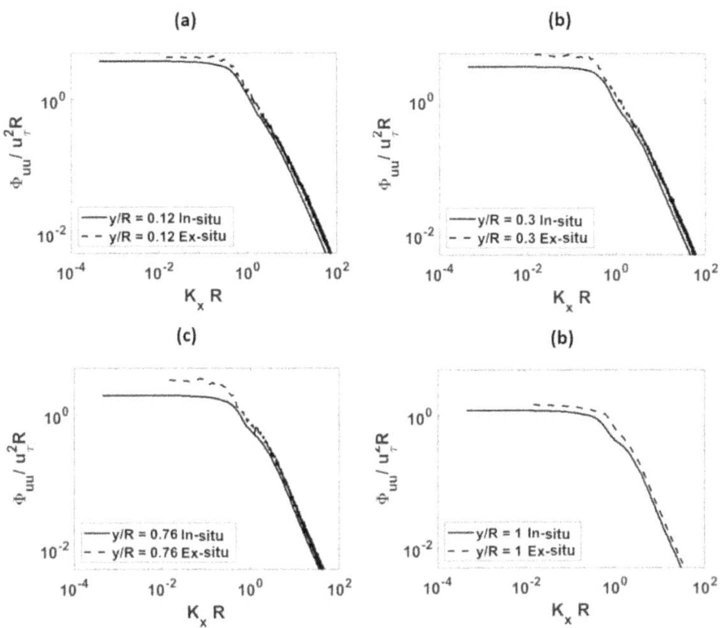

Figure 3.9: Spectra profiles for the In-situ and Ex-situ calibration methods at Re_{τ}6600 (a) y/R=0.12, (b) y/R=0.3, (c) y/R=0.76, and (d) y/R=1.

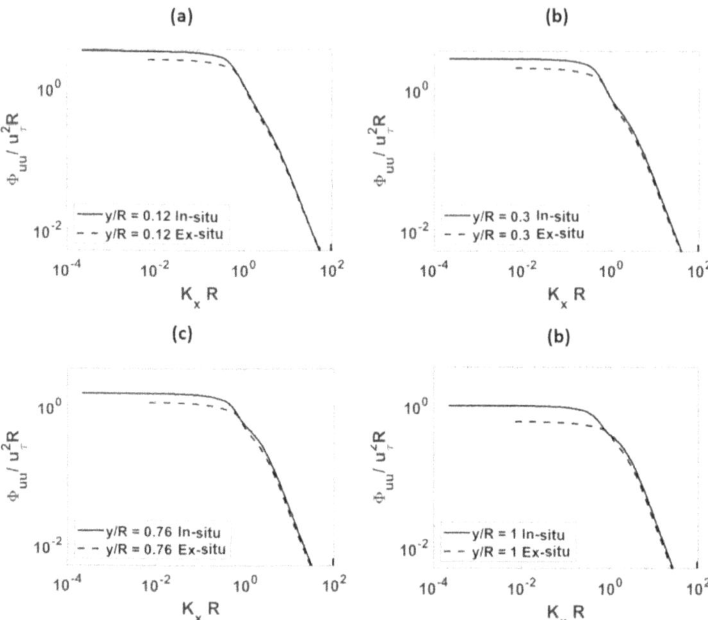

Figure 3.10: Spectra profiles for the In-situ and Ex-situ calibration methods at Re_τ13000 (a) y/R=0.12, (b) y/R=0.3, (c) y/R=0.76, and (d) y/R=1.

The higher Reynolds number shows no significant change in the variation ratio at the four selected wall-normal locations. As shown in fig.(3.10), the lower magnitude of power spectra is obtained from Ex-situ dataset comparing with In-situ results. This behaviour indicates a significant influence of the calibration methods on the lower velocity than higher speeds.

Fig.(3.11) depicts the comparison of pre-multiplied contours at representative wall-normal locations the outer spectral peak is detecting using Vallikivi et al. 2015 and Mathis et al. 2009 estimations. The figures show that although the general shapes of the normalized wavelength at stream-wise velocity component λ_x are similar to the In-situ and Ex-situ calibrations methods. The calculated outer spectral peak is observed to be at different locations at $y^+ \geq 2.4 \times 10^3$ of high Reynolds number and $y^+ \geq 560$ for Re_τ =6600. The Ex-situ outer spectral peak is deviated to approximately 11%, this deviation causes a significant influence on getting an accurate wavelength value.

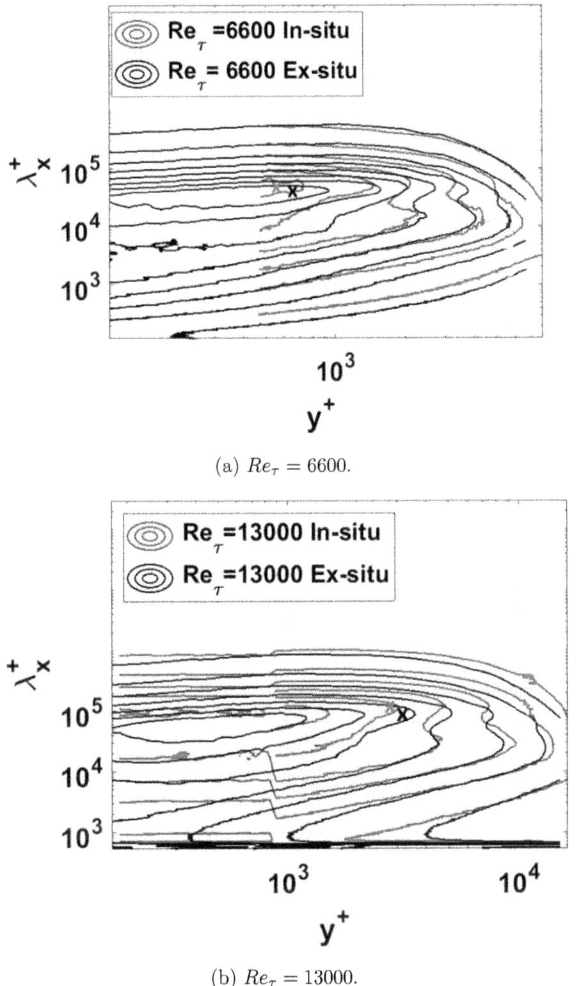

(a) $Re_\tau = 6600$.

(b) $Re_\tau = 13000$.

Figure 3.11: Pre-multiplied spectra contours for the In-situ and Ex-situ calibration methods at different Reynolds number. Black and red symbol (x) represents the outer peak location for the In-situ and Ex-situ respectively.

3.4 Conclusion

The ex-situ calibration method shows a distinguishable effect on hot-wire measurements accuracy observed by the deviation consequence range. This deviation is identified by studying the behaviour of fully turbulent pipe flow via logarithmic velocity profile and statistical moments represented in kurtosis and skewness. Curves clarified that measuring data corresponding to In-situ calibration exhibits satisfactory behaviour compared to Ex-situ calibration data. Although kurtosis profiles for the two cases show a good collapse with DNS data at selected Reynolds numbers. The in-situ method indicates a good agreement with DNS data in skewness than Ex-situ which accomplishes a deviation range of 10-15 $\pm 0.5\%$. The outer peak in pre-multiplied contour spectra is calculated using Vallikivi et al. 2015 and Mathis et al. 2009 estimations. The Ex-situ outer spectral peak showed a deviation of approximately 11% compared to In-situ peak at high Reynolds numbers.

Nevertheless, the behaviour of skewness demonstrates that the deviation of measuring data regarding Ex-situ calibration increases with increasing Reynolds number. The deviation amplitude reduces by rising Re range in the logarithmic velocity profile to reach 2.5 $\pm 0.5\%$ at $Re_\tau = 13000$. It is important to note; Examine the calibration uncertainty through the calibration curve, as fitting the polynomial equation is not sufficient to decide that calibrated hot-wire should perform accurate measurements. This study concluded that HWA In-situ calibrated is preferable to use at low and high velocities.
In-situ calibration has a characteristic feature; it is carried out in the laminar flow field surrounding area that performs more accurate results than the Ex-situ calibration manner. This feature means HWA measurements precision is dependent on the usable calibration method. In future work, more investigations are recommended through using another applicable calibration method, such as calibrating the hot-wire probe at the same location of the measuring station. Comparing calibration methods with each other will assist in explaining the efficiency of each method, as well it proposes which method is convenient for precise hot-wire final results.

Chapter 4: Kinetic Energy Contribution of Coherent Structures in Fully Developed Turbulent Pipe Flow at High Reynolds Numbers

4.1 Experimental Background

Two investigated structures, LSMs, and VLSMs exist in a turbulent pipe flow. These structures have been characterized by visualization, numerical studies, and experiments on different wall-bounded flows (Jiménez 1998 and Kim and Adrian 1999). Coherent structures are commonly used to interpret and understand turbulent physics (Theodorsen, 1952, Robinson, 1991, Jiménez and Moin, 1991, Adrian et al. 2000). Earlier differentiation of such structures has been identified. Guala et al. 2006, Balakumar and Adrian 2007 demonstrated that LSMs are generated by the vortex packets formed when a strain of hairpin structures transport at the same convective velocity. Similar observations have been provided before by Kim and Adrian 1999, Zhou et al. 1999. The LSMs characteristic features are considered in hairpin vortices within the packet align in the stream-wise direction and deduce regions of low-stream-wise momentum between their legs (Brown and Thomas 1977, Adrian et al. 2000, Ganapathisubramani et al. 2003, Tomkins and Adrian 2003, Hutchins et al. 2005). Guala et al. 2006 defined LSMs as structured with the wavelength of 2-3 pipe radii R based on Jiménez 1998 and Kim and Adrian 1999 investigations. On the other side, it is estimated that the VLSMs result from the coherent collection of LSMs in the form of turbulent bulges or packets of hairpin vortices (Kim and Adrian 1999). However, VLSMs in internal flow are recognized such that they have relatively similar characteristic features of LSMs, only being larger in size

(10-20 R) with some characteristic meandering behaviour (Monty et al. 2007). Afzal and Aligarh 1982 proposed that the fully developed turbulent flow in a pipe consists of three layers; outer, intermediate (mesolayer), and inner layer. Very large-scale motions are revealed in the outer layer of fully developed turbulent pipe flow in the shape of high momentum structures of stream-wise velocity fluctuation (Kim and Adrian 1999). The subsequent investigation done by Baily and Smits 2010 exhibits good agreement with Kim and Adrian 1999. They found that VLSMs are observed to sustain well into the outer layer. In the boundary layer, Tutkun et al. 2009 found evidence of weak elongated structures out to the edge of the layer. Otherwise, LSMs have been associated with the occurrence of bulges of turbulent fluid at the edge of the wall layer (Adrian 2007). Evidence for the VLSMs is based on extensive measurements of the turbulent power spectrum of stream-wise velocity fluctuations in a fully developed turbulent pipe flow at high Reynolds numbers (Kim and Adrian 1999).

The existence of energetically considerable large-scale structures in turbulent wall-flow was first recognized by Townsend 1952. Spectral analysis of LSMs and VLSMs indicates that they significantly contribute to the turbulent kinetic energy and Reynolds stress production (Guala et al. 2006). They explained that VLSMs contributed 65% of kinetic energy for scale length greater than 3R and about 35% for scale length 10R. This agreed with Balakumar and Adrian 2007, who demonstrated that such coherent structures carry more than half the kinetic energy and Reynolds shear stress in fully developed pipe flow. For the CoLaPipe, Öngüner 2018 showed that the energy contribution for the structures longer than 3R is 50%, while for the range λ_x /R greater than 10 is 25%. However, it is important to note that at higher Reynolds numbers, the major contribution to the bulk turbulence production comes from the logarithmic region (Smits et al. 2011).

This study is focused on studying the coherent structures in the outer region of fully developed turbulent pipe flow in terms of sizes, scale, and energy contents at specific Reynolds numbers ($Re_\tau = u_\tau R/\nu$)= 2156, 6556, 13132 and 19000, where Re_τ is the Reynolds number based on the friction velocity (u_τ) and the pipe radius (R). Two main objectives are represented in this recent work. First, to complement the previous investigation of the turbulent flow behaviour of LSMs and VLSMs out of the quantitative contribution of total kinetic energy. Second, examining the effect of the CoLaPipe test facility via previous comparative studies of various pipe flow facilities with available data to determine the proper length of large scale structures exploiting the maximum

Reynolds numbers range, $Re_D = u_b D/\nu = 1 \times 10^6$, of the CoLaPipe test facility, where Re_D is the Reynolds number based on the bulk velocity (u_b) and the pipe diameter (D).

4.2 Experimental Setup

The experiments were carried out in Cottbus Large Pipe facility (CoLaPipe) at the Department of Aerodynamics and Fluid Mechanics, Brandenburg University of Technology Cottbus- Senftenberg (BTU C-S), shown in figure (3.1). CoLaPipe is a closed-loop circuit that consists of suction side and return line made of transparent acrylic glass and has an inner diameter of 190 mm and 340 mm, respectively. The facility is provided with a heat exchanger (temperature controlled within $\pm 0.5°$ C) and it allows to explore a wide Reynolds numbers range $5 \times 10^4 \leq Re_D \leq 1 \times 10^6$, where Re_D is the Reynolds number based on the bulk velocity (u_b) and the pipe diameter (D).

Pressure taps are located along the pipe test section. Three pressure taps are installed around the circumference at each pressure measuring location to establish the pressure readings. Current measurements are performed in the suction side at a test section located at a total length to diameter ratio L/D = 110, where the flow is fully developed. In addition, an adequate tripping device (orifice with 10% blockage ratio) is used to ensure the fully developed turbulent flow state. The pressure measurement points $80 < x/D < 100$ are connected to the pressure scanner (PSI9116 Ethernet Pressure Scanner) then static pressure is recorded. The stream-wise pressure gradient dp/dx is obtained by fitting a linear behaviour to determine static pressure measurements. On the other side, the wall friction velocity $u_\tau = \frac{\sqrt{\tau_w}}{\rho}$, that is required for data normalization, is determined from static pressure drop measurements, where ρ is the density of air (working fluid) and τ_w is the wall shear stress estimated as (-D/4)(dp/dx).

Furthermore, the bulk velocity u_b is computed using the pressure drop across the contraction. The friction velocity u_τ and the kinematic viscosity (ν) are used for inner scale normalization, while the outer scale normalization is done by pipe radius R. The acquisition parameters corresponding to each experimental condition are summarized in Table (4.1).

Table 4.1: Experimental flow parameters: Reynolds number (Re_D) based on diameter, Reynolds number (Re_τ) based on friction velocity, centreline (u_c), bulk (u_b), friction (u_τ) velocities, density (ρ), and kinematic viscosity (ν) of the fluid.

	Re_D	Re_τ	u_c [m/s]	u_b [m/s]	u_τ [m/s]	ρ [kg/m^3]	ν [10^{-5} m^2/s]
Case 1	1×10^5	2156	8.91	7.85	0.35	1.19	1.542
Case 2	3.18×10^5	6556	28.48	24.84	1.04	1.19	1.507
Case 3	6.6×10^5	13132	55.36	51.5	2.09	1.21	1.5119
Case 4	1×10^6	19000	79.06	74.96	2.96	1.23	1.525

A standard single boundary layer hot-wire anemometer (Dantec 55P15) is utilized for the four measurement cases. The HWA probe has a diameter (d_w) of 5 μm and (l_w) 1250 μm long, providing length to diameter ratio (l_w/d_w) 250 and viscous scaled wire length range $85 \leq l^+ \leq 247$. It is operated through a Dantec StreamLine Pro frame and 91C10 CTA channels with an overheat ratio $a_w = (R_w - R_o)/R_o = 0.8$, where R_w is the hot-wire resistance at operating temperature and R_o is the resistance at reference (ambient) temperature. The probe calibration (fitted to a 4th order polynomial curve) is performed in-situ through the laminar profile just after the exit contraction at the centreline. The HWA measurements are conducted in a radial direction at the same wall-normal location range $0.05 \leq y/R \leq 1$. The experiments are executed at HWA sampling frequency 10 kHz for $Re_\tau = 2156$, 20 kHz for $Re_\tau = 6556$ and 30 kHz for $Re_\tau = 13132$ and $Re_\tau = 19000$, to record 1.8×10^6, 3.6×10^6, and 5.4×10^6 data samples at each measuring point, respectively. During experiments, the flow temperature is measured via Dantec compact thermistor temperature.

4.3 Validation of Hot-Wire Measurements

It is observed that the large scale structures are convected by the mean flow velocity (u_b), and structures with small scales are extracted from a lower velocity which is highly dependent on the Reynolds number. Del Alamo et al. 2004 suggested that a correct estimation of lower velocities should be considered if the structures in near-wall regions are investigated. Del Alamo's suggestion does not consider in the current study because as mentioned before, the logarithmic region dominates the contribution to bulk production at sufficiently high Reynolds numbers. For that purpose, the investigation of turbulent structures in large scales is executed in the logarithmic and outer region in the present study. Recent experimental hot-wire data are validated by comparing it with other experimental data of SuperPipe test facility at Princeton University (Hultmark et al.

2013 and Morrison et al. 2004) and previous DNS (Ahn et al. 2015) and LES (Chen et al. 2015) numerical datasets. This validation is accomplished via higher-order statistical moments of the stream-wise velocity component at different Reynolds number range, including the present two selected Reynolds numbers $Re_\tau = 6556$ and 19000. The logarithmic velocity profile data in fig.(4.1.a) is normalized by inner scale ($y^+ = yu_\tau/\nu$) to show an acceptable good collapse with Hultmark et al. 2013 experimental data base and Ahn et al. 2015 DNS data. The profile displays the anticipated region of logarithmic dependence given by:

$$u^+ = \frac{1}{k} ln y^+ + B \tag{4.1}$$

The values k=0.4 and B= 4.45 are used throughout the present study proposed by Balley et al. 2014, where k and B are the von Kàrmàn and additive constants, respectively.

On the other side, the variation between Reynolds numbers is noticed in the outer peak of the turbulent intensity representing in the plateau as shows in fig.(4.1.b), similar behaviour is distinctly realized based on the DNS data sets of Ahn et al. 2015. In addition, it is indicated from fig.(4.1.b) that the inner peak for the measuring HWA data is not resolved while the outer peak solely emerges. However, this is defined due to the impact of high Reynolds number on the boundary layer thickness that performs in shrink or collapses the boundary layer towards the wall region.

Otherwise, fig.(4.1.c) and fig.(4.1.d) exhibit the third and fourth statistical moments represented in skewness $\sum_i^N \frac{(u_i - u_m)^3}{N(\delta)^3}$; the measure of the lack of statistical symmetry in the flow, and kurtosis $\sum_i^N \frac{(u_i - u_m)^4}{N(\delta)^4}$; the measure of the amplitude distribution (flatness factor). Where u_i and u_m are the local and mean velocities, respectively. On the other side, δ is the variance $(\sum_i^N \frac{(u_i - u_m)^2}{N-1})^{1/2}$ and N is the number of samples. The validation of the two statistical moments data are done by comparing with Morrison et al. 2004 experimental datasets of Re_τ= 5080 and Chin et al. 2015 LES data. Good compatibility is noticed for skewness as shows in fig.(4.1.c). The calculated kurtosis in fig.(4.1.d) presents equivalent agreement with the earlier proposed data denoted. In skewness, the profile seems to be slightly positive towards the wall region while it starts to change to negative further away from the wall. The dependence of variety Reynolds numbers is noticed, forming a slight deviation in the third and fourth-order moments.

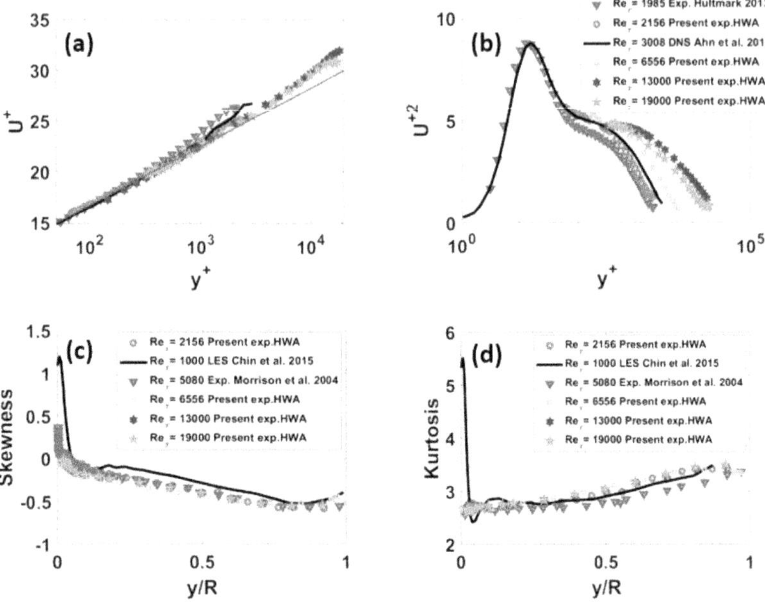

Figure 4.1: Statistical moments of the stream-wise velocity component for the present experimental data compared to experimental and numerical datasets. (a) Mean velocity, (b) Stream-wise Reynolds stress, (c) Skewness, and (d) Kurtosis.

4.4 Results

The Present data are analysed in spectral representation to assess the structures behaviour in the turbulent pipe flow. For identifying the distribution of energy into frequency, the spectrum of the specific signal should be analysed into several frequencies. Noting that the energy distribution in the wavenumber domain helps to understand the behaviour and structures of turbulence. The power spectra and pre-multiplied spectra are commonly used by the turbulence research community, such as Ganapathisubramani et al. 2003, Hutchins and Marusic 2007a, and Perry et al. 1986 who analysed the inner and outer characteristic length scales from the local peaks in the pre-multiplied spectra.

Taylor hypothesis is utilized to enable using wavenumber over the flow direction instead of existing frequency in spectrum calculations (Taylor 1938). Computing the wavenumber from frequency using $k_x = 2\pi f/u_{co}(y)$. The spectrum is computed in wavenumber domain through $\phi_{uu}(k) = P_{uu}(f)u_{co}(y)/2\pi$, where $P_{uu}(f)$ is the power spectra in frequency domain and $u_{co}(y)$ is the convective velocity. In the recent study, bulk mean velocity is carried out as a convective velocity instead of the local mean. This was proposed by Del Alamo and Jiménez 2009 who found that the modes with long wavelengths propagate faster than the local mean velocity in the highly sheared region near the wall.

On the other side, detection of wavelength values in pre-multiplied spectra is corresponding to the identified wavenumber contents of k_x^{-1} and $k_x^{-5/3}$ regions in the power spectrum (Φ_{uu}). k_x^{-1} region divides the inner and outer layer scaling. However, this region is responsible for scaling the sub range of wavenumbers that overlaps the energy-containing (low wavenumber) and attached eddy regions in the spectrum of wall turbulence. In addition, the sub range that overlaps the attached eddy and dissipation (high wave number) regions are expected to scale with $k_x^{-5/3}$ (Smits et al. 2011). Although the $k_x^{-5/3}$ region is well established, at least at a high enough Reynolds number, other experiments indicate that the k_x^{-1} region is only evident at very high Reynolds numbers over a very limited spatial extent (Nickels et al. 2005). The k_x^{-1} region occurs at the four selected Reynolds numbers in the current study.

Figure (4.2) presents the stream-wise velocity power spectra Φ_{uu}, at four Reynolds numbers normalized by friction velocity u_τ and pipe scale R at wall-normal locations range $0.08 \leq y/R \leq 0.7$. The purpose of the three selected Reynolds numbers such as Re_τ = 2156, 6556, and 13132 is to examine the performance of recent measurements data at closer Reynolds number range of other previous accepted experimental data that could enhance the investigation of turbulent flow behaviour at the maximum Reynolds number range ($Re_D = 10^6$) in the CoLaPipe facility. Figure (4.2.b) indicates that behaviour of spectra at $Re_\tau = 6556$ begins and ends at normalized wavenumber range $0.025 \leq k_x R \leq 100$, to appear shorter than behaviours at other Reynolds numbers within the range $0.01 \leq k_x R \leq 100$. The notable phenomenon is estimated to be dependent on the usable signal frequency. Throughout the four cases, the higher spectra behaviours are attained at $y/R < 0.29$, and thence it starts to decrease towards the pipe axis. This interference has been explained earlier by Guala et al. 2006; Vallikivi et al. 2015; and Öngüner 2018. In addition, consistency of energy is obtained towards wall interprets in a merger of wall-normal locations at $0.08 \leq$ y/R ≤ 0.29 for Re_τ = 2156 and 6556 as shown in (4.2.a) and (4.2.b), respectively. At the highest Re_τ= 13132 and 19000, the merger occurs at wall-normal locations y/R = 0.08, 0.17. Although, a salient adhesion of all wall-normal positions is exited at wavenumber $k_x R > 1$, the behaviour within range y/R = 0.08, 0.17, 0.29 at high Re_τ = 19000 starts to separate slightly at $k_x R$ >5. This clarifies well in figures (4.2.c) and (4.2.d), respectively.

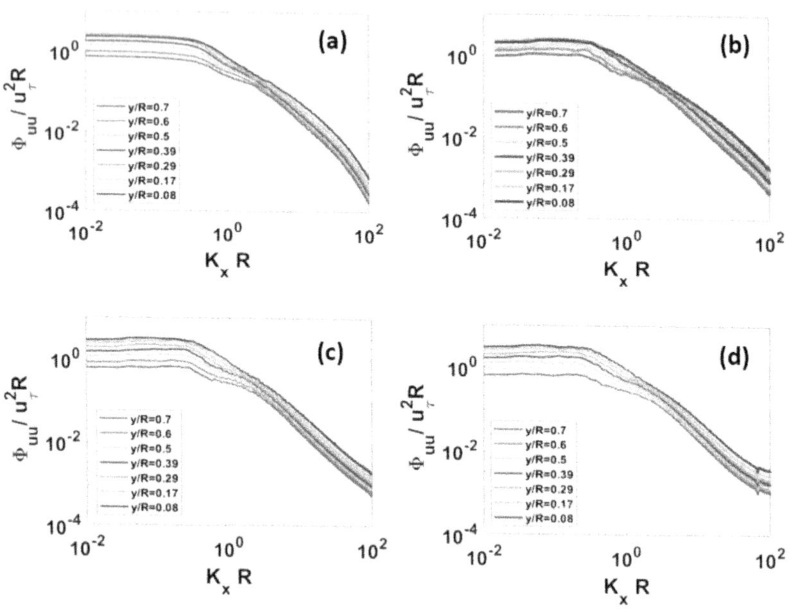

Figure 4.2: Power spectrum of stream-wise velocity fluctuation scaled by pipe radius (R). (a) Re_τ = 2156, (b) Re_τ = 6556, (c) Re_τ = 13132, and (d) Re_τ=19000.

At the same Reynolds number range, the wall-normal scaled power spectra, (inner scaling by (y) exhibits in figure (4.3) to demonstrate k_x^{-1} and $k_x^{-5/3}$ regions. For lowest Re_τ = 2156, the figure indicates that the tendency of k_x^{-1} is revealed at various wall-normal positions relying on Reynolds number. For example, at lowest Reynolds number Re_τ = 2156 the k_x^{-1} region is monitored at $0.08 \leq$ y/R ≤ 0.29 at a normalized wavenumber range $0.64 \leq k_x y \leq 0.8$, while at the highest Reynolds number Re_τ = 19000 a normalized spectra has k_x^{-1} region at y/R= 0.08. The results show that by increasing Reynolds number, the $k_{\overline{x}}^{-1}$ region is well detected but at limited wall-normal locations closest to the wall.

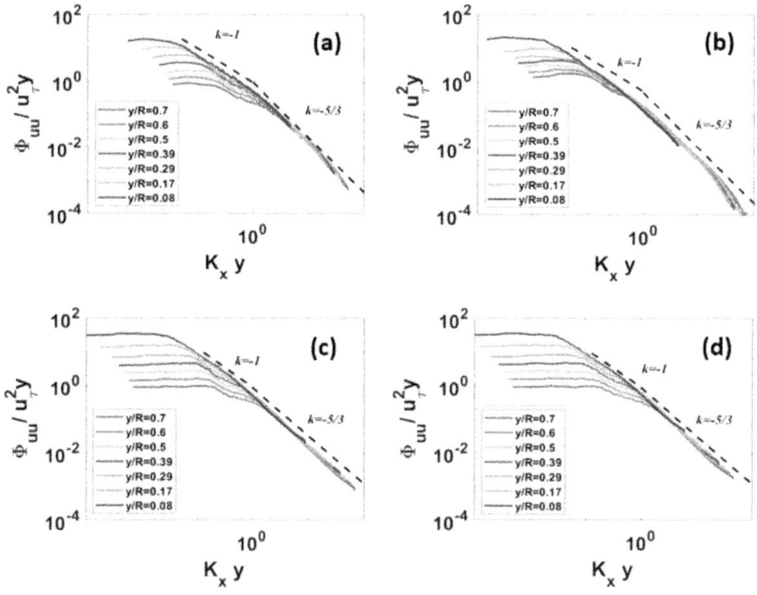

Figure 4.3: Power spectrum of stream-wise velocity fluctuation scaled by wall-normal locations (y). (a) $Re_\tau = 2156$, (b) $Re_\tau = 6556$, (c) $Re_\tau = 13132$, and (d) Re_τ=19000.

Such an expected behaviour is proposed as an influence of high Reynolds number in the boundary layer thickness of pipe flow. In spite of performing the experimental cases at the same measuring location points, each Reynolds number attains a different normalized wall-normal location. For example $Re_\tau = 2156$ and $Re_\tau = 19000$ extend $y^+ = 56, 294$ respectively. Regarding this result, earlier studies established that higher Re conducts a small near-wall region thickness in internal flow (Smits et al. 2011). Moreover, in $k_x^{-5/3}$ region shown in figure(4.3.a) significant full collapse is indicated for the four cases at all wall-normal locations. Nevertheless, the lower Re indicates a coherent collapse, figure (4.3.c) and (4.3.d) demonstrate that such collapse begins to split up with increasing Reynolds number. At $Re_\tau = 19000$, the $k_{\overline{x}}^{-5/3}$ region can be observed at a normalized wavenumber range $1.6 \leq k_x y \leq 6$, while, the tendency of $k_x^{-5/3}$ is perceived within the range $2 \leq k_x y \leq 10$ at $Re_\tau = 2156$. Concluded from spectra behaviours in figures (4.2) and (4.3), that high energy density is deduced in the low wavenumber range (k_x^{-1}) than the higher range ($k_x^{-5/3}$).

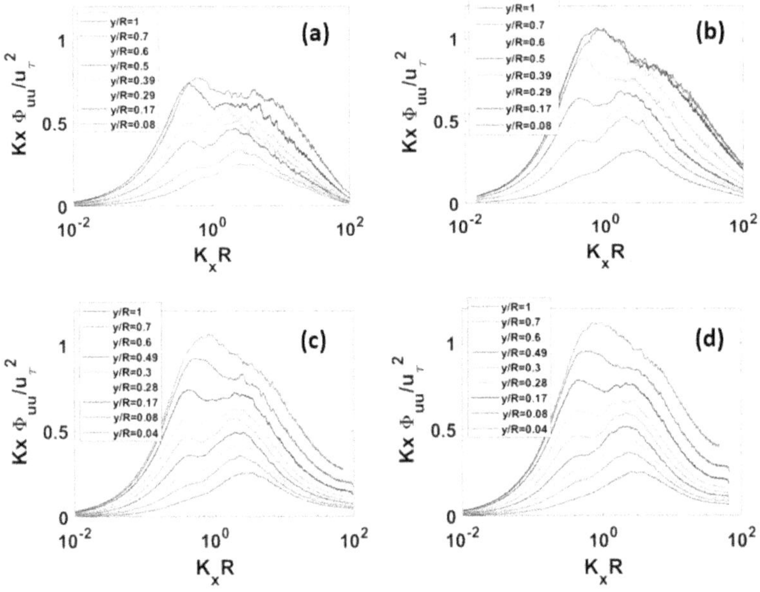

Figure 4.4: Pre-multiplied power spectrum of stream-wise velocity fluctuation (Φ_{uu}/u_τ^2) versus wavenumber scaled by pipe radius ($k_x R$).(a) $Re_\tau = 2156$, (b) $Re_\tau = 6556$, (c) $Re_\tau = 13132$, and (d) Re_τ=19000.

Pre-multiplied spectra are used to classify the turbulent structures regarding their wavelengths via two major distinguishable peaks; the lower and the higher wavenumber peaks are represented in very large and large scale motions, respectively. It is worth noting that the inner peak cannot be easily identified because all measurements are conducted in the logarithmic and outer regions $56 \leq y^+ \leq 18000$ at high Reynolds numbers as shown in fig.(4.1.b). It is proposed to take the most energetic peak near the wall region outside the viscous region as an inner peak (Vallikivi et al. 2015). The evolution of the two distinct energy peaks in pre-multiplied stream-wise velocity spectra are shown in figure (4.4) at the four selected Reynolds numbers.

It is clearly noticed in figure (4.4.a), that inner peaks are recognized at all wall-normal locations, while the outer peaks are detected obviously at $0.08 \leq y/R \leq 0.5$. Otherwise, figures(4.4.b,c,d) present the detection of outer peaks than inner peaks at higher Reynolds numbers. However, it is hard to distinguish the high wavenumber peak at some radial locations by increasing the Reynolds number in the pre-multiplied spectrum. Such inference is expected as well, and it has been proved earlier by Kim and Adrian 1999.

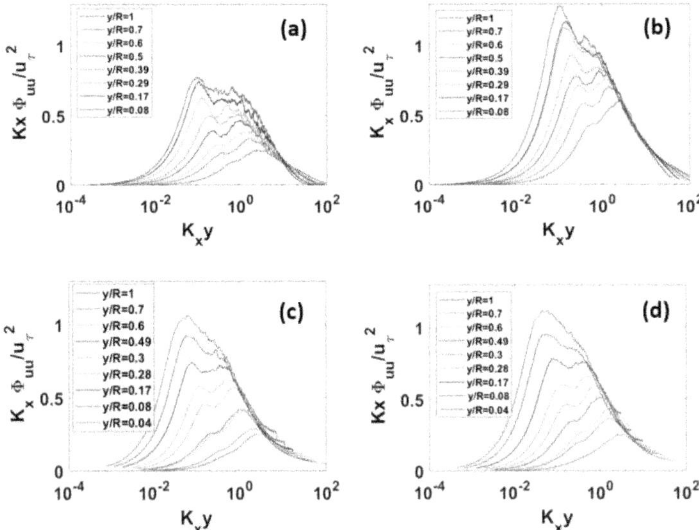

Figure 4.5: Pre-multiplied power spectrum of stream-wise velocity fluctuation (Φ_{uu}/u_{τ}^2) versus wavenumber scaled by wall-normal locations ($k_x y$). (a) $Re_\tau = 2156$, (b) $Re_\tau = 6556$, (c) $Re_\tau = 13132$, and (d) Re_τ=19000.

Pre-multiplied spectra normalized by inner scaling y in figure (4.5) indicates the same behaviour in figure (4.4), excluding the noticeable collapse of all wall-normal locations at the four Reynolds numbers. By increasing the Reynolds number, the collapse increases to be fully cohesion at $Re_\tau = 19000$. The same observation is demonstrated by Vallikivi et al. 2015. Also, outer peaks are monitored to be slightly faded in the distance from the wall, and the inner peaks are hidden due to collapse formation in the high wavenumber region. Mainly distinct in both figures (4.5.a) and (4.5.d), that spectra at $Re_\tau = 2156$

attain less energy comparing with higher Reynolds numbers to accomplish its maximum values at $Re_\tau = 19000$.

Based on the computed wavelength in the pre-multiplied spectra, the outer spectral peak location (y^*_{osp}) for very large scale motions can be identified using Mathis et al. 2009 estimation. Their estimation of y^+_{osp} in pipe is used in a dependence of Reynolds number, $y^+_{osp} \approx 3.9\sqrt{Re_\tau}$. Here in the present study, the outer spectral peak is revealed in the logarithmic layer for Re_τ =2156 at $y^+ \approx 56$ as shown in the contour plots in fig.(4.6). As depicted in fig.(4.6.a), the y^+_{osp} appears to follow the turbulent intensity outer peak, which has been observed previously by Vallikivi et al. 2015.

The overall spectral peak behaviour is similar at the two Reynolds numbers except for a slight increase of y^+ value at $Re_\tau = 19000$ that appears in the plateau region of y^+_{osp}. This behaviour agrees with Hutchins et al. 2007 and Smits et al. 2011, who demonstrated that the outer spectral peak in turbulent intensity starts only to be constant at frictional Reynolds number $Re_\tau > 20000$.

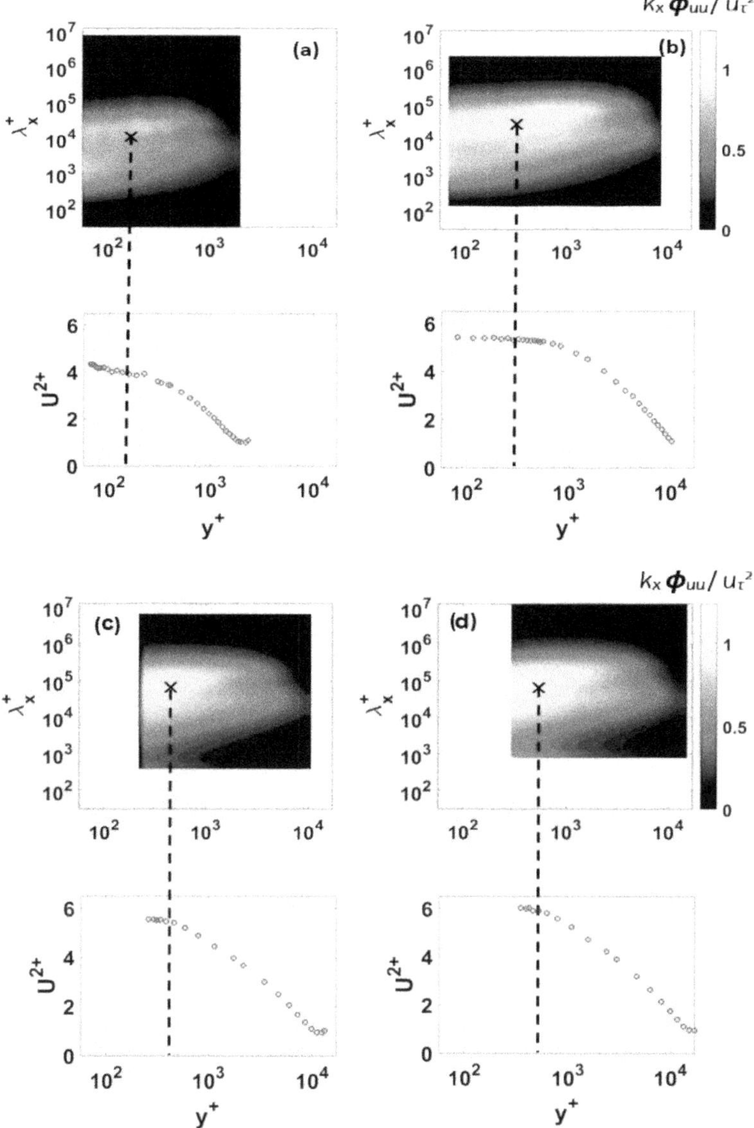

Figure 4.6: Upper row: Contour plots of spectra ($k_x\Phi_{uu}/u_\tau^2$). Bottom row: turbulent intensity (u^{2+}) profiles. (a) $Re_\tau = 2156$, (b) $Re_\tau = 6556$, (c) $Re_\tau = 13132$, and (d) Re_τ=19000. Symbol (X): location of the outer spectral peak y^+_{osp}.

The absence of the inner spectral peak in the contour and turbulent intensity plot is regarding to the physical extent of the thin wall layer due to the applied high Reynolds number as mentioned previously in the first section. Besides, the VLSMs peaks exist at λ_x > 3R to state the largest energy content outside the viscous wall region corresponding to the logarithmic outer layer. This present in outer scaled spectra as in fig.(4.7).

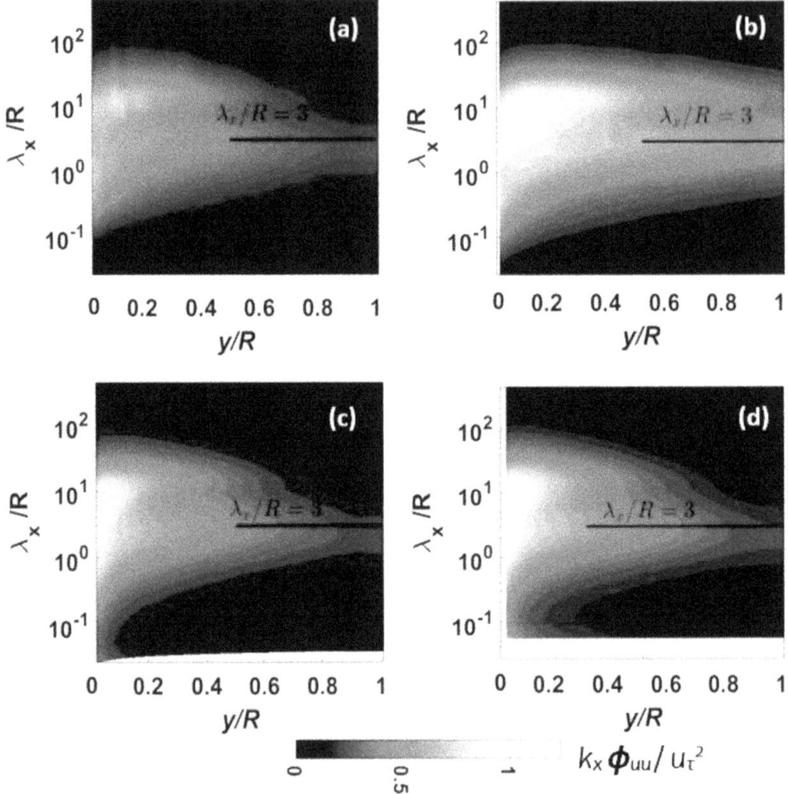

Figure 4.7: Contour plots of spectra ($k_x\Phi_{uu}/u_\tau^2$) in outer scaling. (a) $Re_\tau = 2156$, (b) $Re_\tau = 6556$, (c) $Re_\tau = 13132$, and (d) Re_τ=19000.

The wavelength of peaks in the one-dimensional pre-multiplied energy spectra is interpreted in figure (4.8). To denote that results agree with the behaviours and values found in other experiments, figure (4.8) shows scattered data for selected results of previous pipe flow experiments (e.g. Center for International Cooperation in Long Pipe Experiments (CICLoPE) and CoLaPipe) (Öngüner et al. 2017a; Öngüner 2018) compared to the present one. According to the comparison of the entire experimental data, including the present experiment, it is noticed that the maximum wavelength scale values for large and very large scale structures are realized in the outer layer at half pipe radius y/R =0.5 for all data. The minimum values are observed in the direction towards the wall. The wavelength values (λ_{max}) of VLSMs corresponding to outer peaks in pre-multiplied spectra can be detected at wall-normal locations y/R $\leq$ 0.5 with a maximum value (λ_{max})= 19 R for Re_τ=19000. Subsequently, the behaviour of VLSMs results is apparent to completely vanish at y/R > 0.5.

On the other side, recent data presents no significant difference in the wavelength values of LSMs to acquire 3 R as a maximum value at Re_τ=19000. This phenomenon exhibits a good agreement with Guala et al. 2006, Kim and Adrian 1999, Bullock et al. 1978 and Perry and Abell 1975; Balakumar and Adrian 2007 explanations but not sufficient agreement with Zanoun et al. 2017 and Öngüner 2018 observations, who indicated that the VLSM submerged in LSM at wall-normal locations greater than 0.5. Another perceived observation agrees with Guala et al. 2006, represented in figure (4.8), by increasing Reynolds number, the wavelength value of VLSMs is increased slowly.

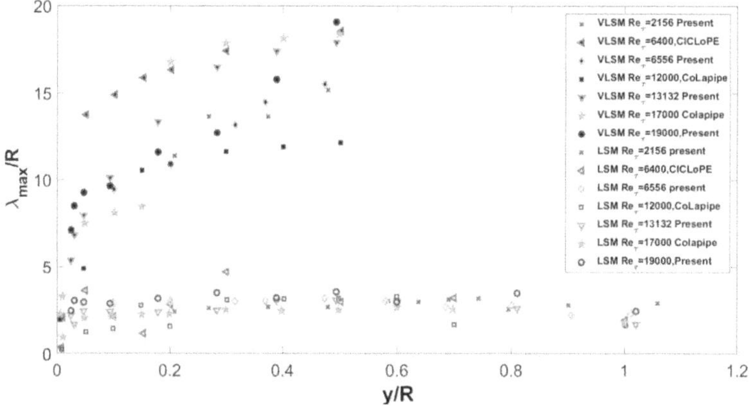

Figure 4.8: Wavelengths of the peaks in the pre-multiplied power spectra of stream-wise velocity.

Contribution of large and very large scales to kinetic energy can be estimated by examining the cumulative contribution of all wavenumbers from $k_x = 2\pi/\lambda$ to infinity,

$$\gamma_{ij}\left(k_x = \frac{2\pi}{\lambda_x}\right) = 1 - \frac{\sum_0^{k_x} \phi_{ij}(k_x)}{\sum_0^{k_x max} \phi_{ij}(k_x)} \tag{4.2}$$

γ_{ij} is defined as a cumulative contribution for all wavelengths from λ_x to 0. Figures (4.9) and (4.10) show the cumulative energy fraction, γ_{uu} at wall-normal locations 0.048 $\leq$ y/R$\leq$ 0.7 for each measurement case. Corresponding to the wavelength values of LSMs and VLSMs that are quantified from spectral plots and presented in figure (4.8), the widely used separation scale λ_x = 3R has relied as a reference in figures (4.9) and (4.10), and as represented the vertical dashed line to separate the dominated LSMs and evaluates VLSMs.

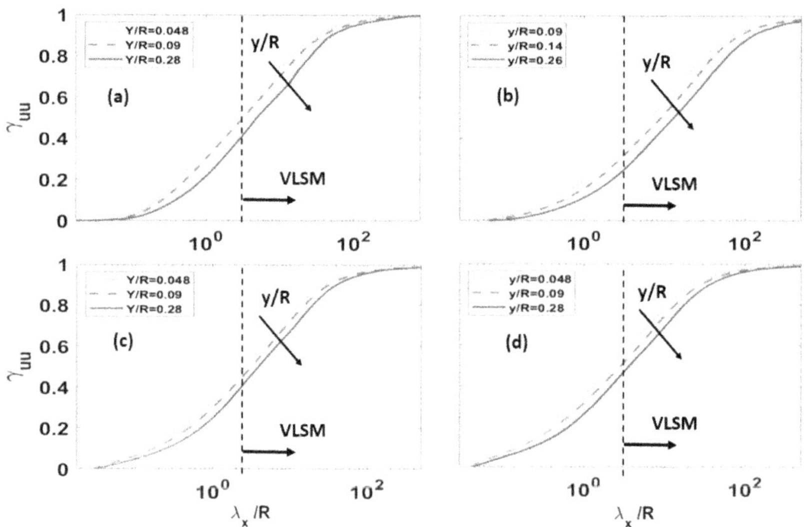

Figure 4.9: Cumulative stream-wise kinetic energy fraction (γ_{uu}) for 0.048 $\leq$ y/R $\leq$ 0.28. (a) Re_τ= 2156,(b) Re_τ= 6556, (c) Re_τ= 13132, and (d) Re_τ= 19000.Vertical dashed lines refer to the position of $\lambda_x/R = 3$.

It is clearly seen that the behaviour for all measurement cases is similar. Also it is clear that at all Reynolds numbers, cumulative energy curve increases towards near-wall region within the range 0.048 $\leq$ y/R$\leq$ 0.28, then it starts to decrease across pipe axis at wall-normal location y/R > 0.28. This phenomenon can be clarified entirely through two separate plots in figures (4.9) and (4.10) referring to near wall and pipe axis regions, respectively.

Such phenomena are consistent with the cumulative energy results explained by Guala et al. 2006 as well as it agrees with Monty et al. 2009 consideration who found that the VLSMs in pipe and channel flows move to longer wavelengths with distance from the wall. In contrast, the opposite occurs in external flow, in which VLSMs dissipate suddenly beyond the log region (Smits et al. 2011). Monty et al. 2009 explanation is based on earlier investigations done on large-scale phenomena by Hutchins and Marusic 2007a and Kim and Adrian 1999. Regarding to figures (4.9) and (4.10), it is deduced that energy contribution of very large scale structures greater than 3R is about 55 %, while structures with wavelength λ_x > 10 R contain 25 − 30%. The current results show apparent agreement with Guala et al. 2006 estimations in pipe flow and Balakumar and Adrian 2007 in the boundary layer flow.

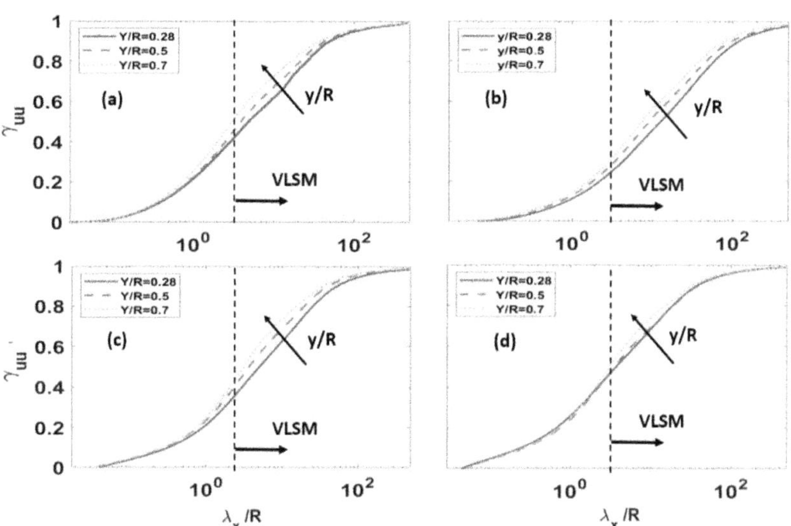

Figure 4.10: Cumulative stream-wise kinetic energy fraction (γ_{uu}) for 0.28 ≤ y/R ≤ 0.7. (a) Re_τ= 2156,(b) Re_τ= 6556, (c) Re_τ= 13132, and (d) Re_τ= 19000.Vertical dashed lines refer to the position of λ_x/R=3.

4.5 Conclusion

The results of the spectral analysis of stream-wise velocity at $Re_\tau = 2156$, 6556, 13132, and 19000 in the present study revealed similar recognized characteristic features of large scale motions (LSMs) in fully developed turbulent pipe flow. It is considered that the four different high Reynolds numbers did not induce any significant influence on the wavelength value of LSMs to exceed 3R as explained previously by Guala et al. 2006, Jiménez et al. 1998 and Kim and Adrian 1999. Otherwise, the very large scale motions (VLSMs) present a slowly increase in their wavelength value to attain 19R as an extreme value for the maximum Reynolds number range $Re_\tau = 19000$ in the CoLaPipe facility. Throughout the power spectra, the k_x^{-1} slope is detected in the four experimental cases at low wavenumber region $k_x < 1$, while at higher Reynolds numbers $Re_\tau = 13132$ and 19000, the k_x^{-1} region is only evident over a very limited spatial extent y/R = 0.08 due to the elimination of boundary layer thickness. This inference is consistent with Nickels et al. 2005 investigations.
Furthermore, outer peaks in pre-multiplied spectra are monitored at radial locations close to the wall and roughly distinguishable towards the pipe axis with increasing Reynolds number. Whilst it is well established that inner spectral peaks are hardly distinct at high Reynolds numbers than the lowest range $Re_\tau = 2156$ performing the high agreement with Kim and Adrian 1999 explanations. The location of the outer spectral peak (y^+_{osp}) is detected by using Mathis et al. 2009 estimation to be identified at normalized wall-normal location $y^+ \approx 56$ for $Re_\tau = 2156$ in the contour plots, and its location increases with increasing Reynolds number. The y^+_{osp} in pre-multiplied spectra contours is observed to follow the outer spectral peak in turbulent intensity, which has been demonstrated earlier by Vallikivi et al. 2015. The outer peaks are obtained at wavelength $\lambda_x > 3R$ to present the largest energy content outside the viscous wall region.

The cumulative energy fraction estimation is identified based on the wavelength value of very large and large scale motions corresponding to the low and high wavenumber peaks in the one-dimensional pre-multiplied power spectra, respectively. Consequently, it is realized that VLSMs with wavelength value greater than 3R carry about 55% of the energy of the stream-wise velocity component in the outer layer y/R > 0.28 (the most energetic region). Moreover, structures longer than 10R contribute 25-30 % of total kinetic energy. Cumulative energy results in the current study expose sufficient agreement with previous studies Guala et al. 2006 and Balakumar and Adrian 2007.

Visualization studies are recommended to enhance investigations of coherent structures, particularly VLSMs in turbulent pipe flow. Even it is difficult to observe the very large scale motions by visualization or by Particle Image Velocimetry (PIV) due to their great size (Guala et al. 2006). Still, PIV is widely used to perform flow field measurements in the stream-wise wall-normal plane.

Chapter 5: PIV Measurements in Pipe Flow

5.1 Introduction

Earlier experimental and numerical studies have explored and documented the existence of coherent structures among the investigation of hairpin-like vortices inclined at an angle to the free stream Panton 1997. Subsequent studies demonstrated that multiple structures of these hairpins are travelled at the same convective velocity forming vortex packets that create large scale structures in internal and external flows Kim and Adrian 1999, Balakumar and Adrian 2007. The features of large structures are their alignment in the stream-wise direction to produce regions of low and high stream-wise momentums in the inner and outer layers Hutchins et al. 2005. Large coherent structures are known to inhabit the log and outer regions of the boundary layer at moderate to high Reynolds number Hutchins and Marusic 2007a to be responsible for ejection and sweep events generation, which are the major contributors to turbulent production and Reynolds shear stress (Martins et al. 2019). High-speed PIV measurements are performed for flow visualisation to investigate the behaviour of large scale structures in x-y planes.

5.2 Experimental Setup

Two-dimensional planar PIV measurements are conducted in Brandenburg University of Technology pipe test facility (CoLaPipe) with a diameter of 190 mm. The measurement station is located at length to diameter ratio L/D =110, shown in fig.(5.1). A glass segment test section of two meters long is mounted instead of plexiglass for clear optical accesses. An acrylic self-adhesive black foil is suited to an area 1/3 of the pipe diameter to eliminate background laser light reflection. The experimental setup shows a high standard Nd: YLF double-pulse laser with frequency 1.4kHz and wavelength 532 nm per pulse placed with a view angle of about 45° to face the uncoated transparent area of the test section.

A phantom VEO 640L CMOS 16-bit camera with 2560 ×1600 pixels locates vertically underneath the pipe to be perpendicular to the laser sheet. The temporal distance between two consecutive sets of velocity snapshots is 0.25 t_c as suggested by Benedict and Gould 1996, with t_c corresponding to the convective time scale (2 R/u_b) based on the pipe radius (R) and bulk velocity (u_b). Therefore, acquisition frequency (f) is adjected to be 480 Hz for Re_τ =3200 and 799 Hz for the two higher measuring Reynolds numbers. Table (1) presents the flow parameters during the recent experimental investigation. The pressure gradient is measured upstream and at the exit nozzle contraction of the pipe by utilizing the venturi effect to calculate the bulk velocity.

Table 5.1: Experimental flow parameters. Re_D is Reynolds number based on pipe diameter and bulk velocity (u_b), while Re_τ is Reynolds number based on pipe radii and friction velocity (u_τ)

Case	Re_D	Re_τ	u_b [m/s]	u_τ [m/s]	ρ $[kg/m^3]$	ν $[m^2/s]$
1	1.45×10^5	3200	11.7	0.52	1.19	1.53×10^{-5}
2	3.18×10^5	6605	25.3	1.04	1.19	1.52×10^{-5}
3	5.33×10^5	10617	42.2	1.58	1.22	1.49×10^{-5}

An optical calibration is performed using a two-dimensional calibration plate with dimensions (100 × 200 mm) prepared in the department's laboratory. This calibration plate is designed with 3 mm circular patterns and a distance of 2 mm between each point. A high seeding generator is used to inject a small liquid of standard DEHS (Di-Ethyl-Hexyl-Sabcat) with an average diameter of $< 2\mu$m fixed before the inlet of the settling chamber of the CoLaPipe test facility. Subsequent 1838 snapshots are efficiently acquired.

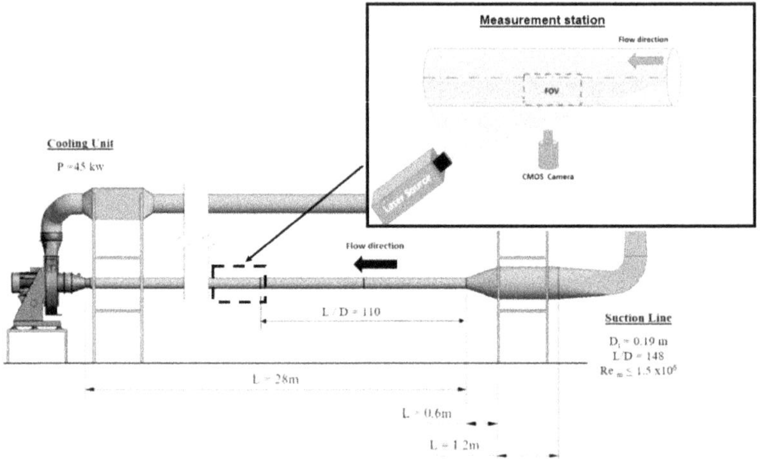

Figure 5.1: PIV setup in CoLaPipe test facility at LAS.

5.3 PIV Measurements Validation

The PIV measurements are validated through the four statistical moments obtained from the current measurements, earlier Direct Numerical Simulations (DNS), Large eddies simulations (LES), and experimental data sets. The converged velocity statistics of the PIV database are acquired from 1838 instantaneous velocity snapshots at three relatively high Reynolds numbers. In the recent study, comparisons are accomplished across the entire wall-normal extent of the boundary layer for x-y plane datasets. Herein, the extended statistics are computed from the average of the field of view (FOV) after excluding noisy edges. However, the mean flow field is $0 \leq x/R \leq 2$, but due to the pipe reflection, the borders of FOV are reduced to be $0.4 \leq x/R \leq 1.6$. Fig.(5.2a & b), exhibit the mean stream-wise velocity profile (u^+), the stream-wise and wall-normal turbulence intensities (u^{2+}) and (v^{2+}), and the Reynolds shear stress ($-uv^+$) for the present PIV database at Re_τ= 3200, 6565, and 10600. For comparison, Ahn et al. 2015 (DNS data) statistical results and Zagarola and Smits 1998 experimental datasets are involved at Reynolds numbers Re_τ= 3008, Re_τ= 10480, respectively. Recent HWA dataset Re_τ= 6556, presented in chapter(4) is also included.

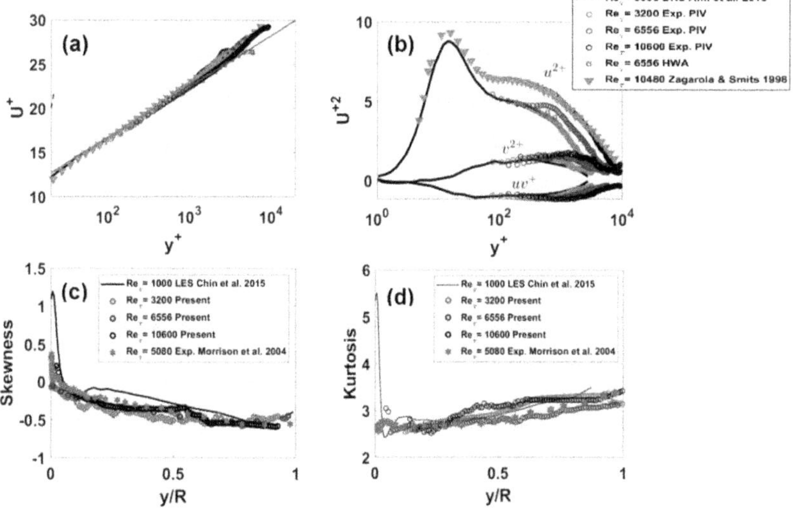

Figure 5.2: Comparison between the present PIV experimental datasets versus DNS datasets at $Re_\tau = 3008$ (Ahn et al. 2015), LES dataset $Re_\tau = 1000$ (Chin et al. 2015), $Re_\tau = 10480$ (Zagarola and Smits 1998), $Re_\tau = 5080$ (Morrison et al. 2004), and $Re_\tau = 6556$ (HWA measurements). (a) Logarithmic velocity profile (u^+), (b) Stream-wise and wall-normal turbulent Intensity (u^{2+} and v^{2+}) and Reynolds Shear stress ($-uv^+$), (c) Skewness, and (d) Kurtosis.

The DNS datasets are selected to be at matched Re_τ= 3200 of the PIV database. This validation provides convenient means to benchmark the quality of the experiment, segregating any experimental uncertainties which are not found in DNS. The first and second flow statistics show a significant agreement between PIV and DNS dataset. In addition, the data shows a significant agreement with the equivalent Reynolds number data of hot-wire measurement at $Re_\tau = 6556$. The influence of spatial attenuation and wall reflections causes large discrepancies, particularly in the near-wall region due to wall turbulence. For that, the near-wall region y^+ <60 is not clearly captured using the PIV system. The PIV data at higher Reynolds number is strongly matched with Zagarola and Smits 1998 hot-wire data.

The skewness ($S = \langle u\rangle^3/\langle u^2\rangle^{3/2}$) is presented in fig.(5.2.c). The flow of the PIV dataset demonstrated a similar close behaviour with the statistical LES dataset Chin et al. 2015 and hot-wire anemometry dataset of Morrison et al. 2004, the skewness of the current data agrees with Morrison results in the outer coordinates at region y/R $\geq$ 0.6, while in the inner coordinates, a slight deviation is observed at 0.1 $\leq$ y/R $\leq$ 0.5. These variances are due to the dependence of variety Reynolds numbers (Hutchins et al. 2009). The skewness profile is apparently positive toward the wall region, and it changes to negative further from the wall. As well, a variation between the PIV database and LES is nearly noticed at 0.2 $\leq$ y/R$\leq$ 0.6.

Fig.(5.2.d) shows the kurtosis ($K = \langle u^4\rangle/\langle u^2\rangle^2$), an acceptable good agreement is acquired, However, kurtosis behaviour obtained from PIV data is realised to be roughly similar to the LES database and Morrison et al. 2004 experimental results as presented previously for skewness in fig.(5.2.c).

5.3.1 Validation of Spectral Analysis

For the simplicity of description in this section, u_i (i=1,2) was adopted to denote the stream-wise and the wall-normal fluctuating velocity signals with u_1= u and u_2= v. Then the power spectrum or (cross-power spectrum) can be defined as :

$$\Phi_{u_iu_j}(f) = c \mid F_{u_i}(f)F^*_{u_j}(f) \mid, \tag{5.1}$$

where $F_{u_i}(f)$ denotes the Fourier transformation of the fluctuating velocity signal u_i, f is the frequency, superscript * indicates the complex conjugate, | | designates the modules, and c is a constant determined by the satisfying following equation :

$$\overline{u_iu_j} = \int_o^\infty \Phi_{u_iu_j}(f)df \tag{5.2}$$

Based on Taylors's frozen turbulence hypothesis (Taylor 1938), the frequency-based spectrum $\Phi_{u_1u_1}$ can be transformed to stream-wise wavenumber (k_x)/ wavelength (λ_x) spectrum $\Phi_{u_1u_1}(k_x)$ /$\Phi_{u_1u_1}(\lambda_x)$ via the following equation:

$$k_x = \frac{2\pi f}{u_1(y)}, \quad or \quad \lambda_x = \frac{2\pi}{k_x} \tag{5.3}$$

Where u_1(y) is the local mean stream-wise velocity at y. Even though Taylor's hypothesis may not be accurate in the low wavenumber range, the errors involved in Taylor's hypothesis will not corrupt the recent results (Guala et al. 2006).

Welch's overlapped segments averaging method is used to obtain the wavelength spectrum for the velocity signals at each wall-normal location. It is given that the raw spectrum derived from a raw velocity signal would be perversive by the high-frequency measurement noise. A smooth fraction of 40% is applied to the raw velocity signals. Smoothing the spectrum reduces the noise effects and facilitates identifying the original peaks of large and very large scale motions, without altering the shape of the spectrum.

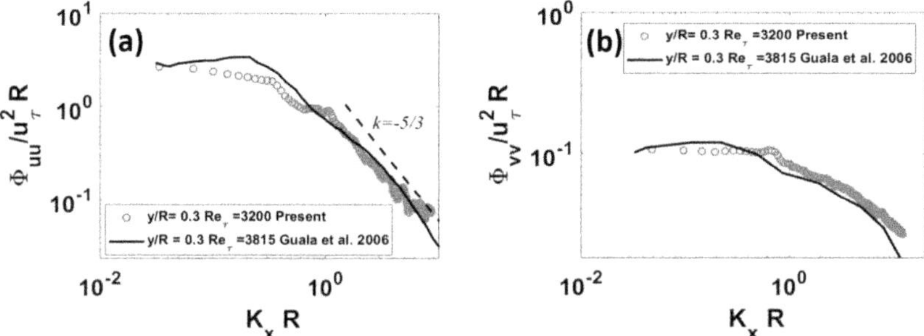

Figure 5.3: Wavenumber spectrum velocity fluctuation for the present PIV datasets ($Re_\tau = 3200$) versus other experimental pipe flow ($Re_\tau = 3815$) (Guala et al. 2006) at y/R=0.3. (a) Power spectra for stream-wise velocity fluctuation, and (b) Power spectra for wall-normal velocity fluctuation.

Fig.(5.3) presents the filtered power spectrum for stream-wise and wall-normal velocity components at y/R = 0.3. For comparison, the spectral reported results of other experimental pipe flow (Gaula et al. 2006) are included at a similar y location and closely comparable Re_τ. Slight measurement noise is observed for the PIV data at the high wavenumber region, that almost interpret as a rapid increase in energy at the small wavelength (Atkinson et al. 2014).

Although only one wall-normal location is shown here for u and v velocity components, a distinguishable match between the PIV data and other hot-wire anemometry datasets is indicated from the monitored results across the entire boundary layer. Based on the comparison, good agreement verifies the measurements accuracy and the rationality of the post-processing method.

5.4 Characterization of Turbulent Structures

One of the most challenging applications of particle image velocimetry (PIV) is the measurement of turbulence. For example, the measurement in turbulent pipe flows (i.e. pipe and channel flows) may require an image size large enough to include the flow at the pipe radii with a sufficient resolution to resolve the small scale motions in a stream-wise direction. Herein, time-resolved PIV measurements are performed to measure the stream-wise/wall-normal velocity fluctuations in the overlap and outer region with an effective spatial resolution and measurement accuracy to resolve the structures of turbulent flows. A large number of snapshots are captured at a wide range of Reynolds number $1663 \leq Re_{\tau} \leq 10617$ to determine the flow statistics and to investigate the dynamics of coherent structures. Fig.(5.4) presents the outer scaled mean stream-wise velocity contour normalized by corresponding mean velocity at the pipe axis (u_b/u_c). Velocity variation can be observed across the flow field at three selected different Reynolds numbers to indicate the same performance for all measured Reynolds numbers.

Furthermore, large fluctuation of stream-wise component (u^+_{rms}) depicts small magnitude in the core region and larger in the near-wall region, this is clarified well in the inner scaled u_{rms}/u_{τ} contours as in fig.(5.5). The conveniences of velocity fluctuation, known as Reynolds shear stress, appear as an additional term in the Reynolds averaged Navier-Stokes equations. However, they play an apparent role in turbulent stresses representing in the uv velocity components. Reynolds shear stress is recognized to be settled in the near-wall region and decreases towards the pipe axis. This explanation done by previous works of literature are converged with the accomplished present results exhibit through Reynolds shear stress (uv^+) contours in fig.(5.6).

The current measurements' temporal and spatial resolutions enable us to visualize large structures known to exist in the near-wall region at moderate to high Reynolds number in pipe flow. Fig(5.7) shows a typical result of subsequently six instantaneous stream-wise velocity fluctuations snapshots measured in fully developed turbulent pipe flow of total captured snapshots 1838 at $u_b = 11.7$ m/s. Total snapshot of PIV data in stream-wise/wall-normal plane has an acquisition time of about 3.05 s and a time variance of 2 ms between each two consequence images. The bottom and top axis in the figure coincide with the pipe wall and pipe axis, respectively.

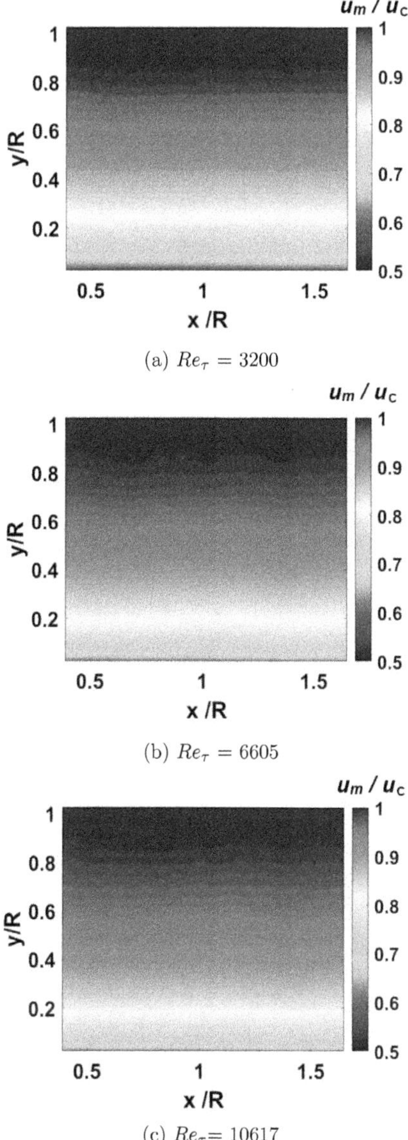

(a) Re_τ = 3200

(b) Re_τ = 6605

(c) Re_τ= 10617

Figure 5.4: Contour plots of mean stream-wise velocity (u_m) normalised by corresponding mean velocity at pipe axis (u_c).

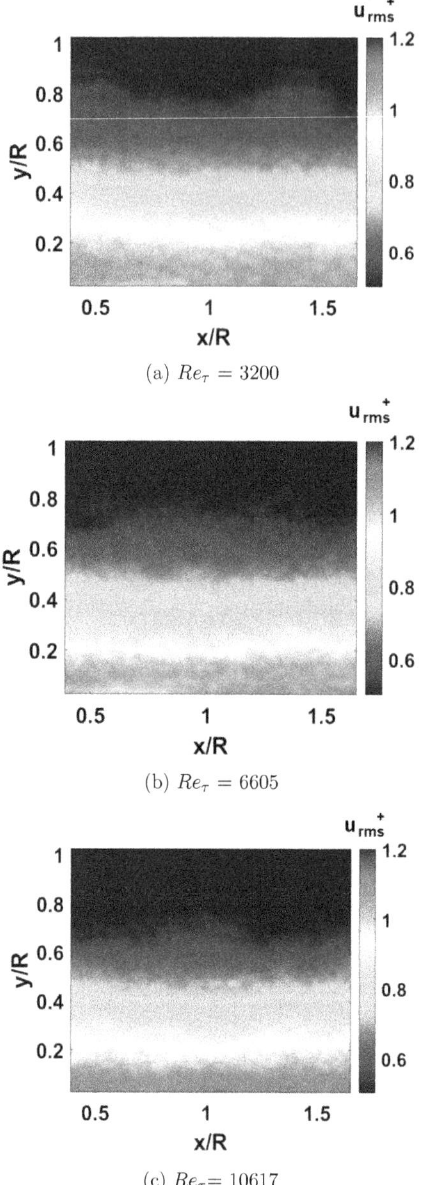

(a) $Re_\tau = 3200$

(b) $Re_\tau = 6605$

(c) Re_τ= 10617

Figure 5.5: Contour plots of stream-wise (u_{rms}) velocity normalised by friction velocity (u_τ).

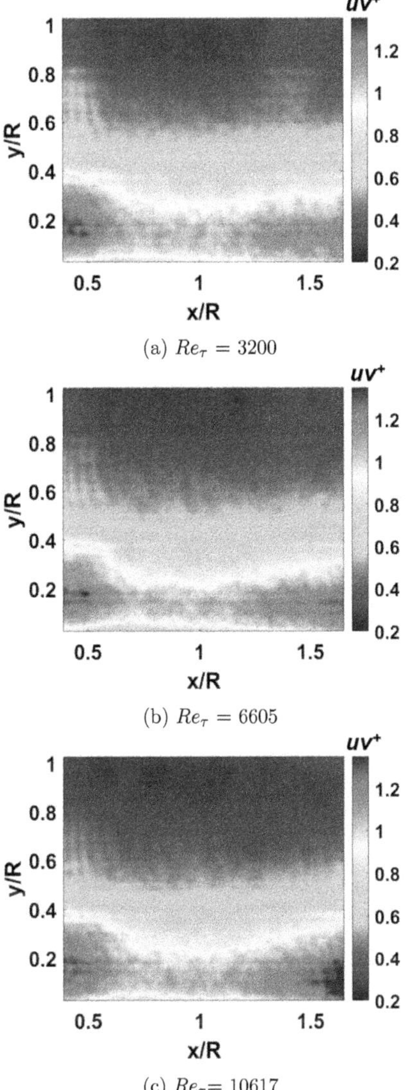

(a) $Re_\tau = 3200$

(b) $Re_\tau = 6605$

(c) $Re_\tau = 10617$

Figure 5.6: Contour plots of Reynolds shear stress velocity (uv) normalised by friction velocity (u_τ).

The images provide several hairpin packets in the logarithmic region characterized by spatially coherent structures with a range of different scale packets coexistence. This observation explains that the spacing of low-speed streaks in the stream-wise velocity field increases across the logarithmic region with distance from the wall (Ganapathisubramani et al. 2003; 2005; Tomkins and Adrian 2005).

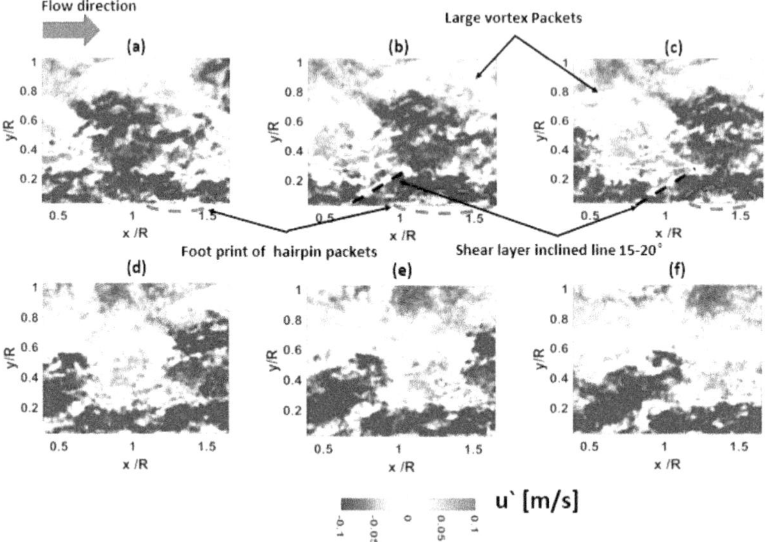

Figure 5.7: Sequential snapshots of PIV instantaneous fluctuations velocity fields with equivalent time interval between each other Δ t=0.002s at Re_τ= 3200.

In addition, the hairpin group forming large and very large structures are noticed to be inclined at a characteristic angle of 15-20° to the wall. These inclined structures are defined to be corresponding to the shear layer. However, the direct experimental observations of Liu et al. 1991 explained that regions containing high Reynolds stress in the flow associated with these near-wall shear layers. Furthermore, Head and Bandyopadhyay 1981, who studied speed time-sequenced images of the smoke-filled boundary layer, concluded that the hairpin packets form an inclined angle of 45° to the wall. The inclined shear layer separated between low speed-streaks of the low-speed fluid in the near-wall region (close to the wall) and the higher speed fluid in the logarithmic region in the outer layer. The inclined shear layer is caused when the second quadrant ejections of the low-speed fluid move through the inclined loop of the hairpin vortex-induced from the legs and the head to encounter a Q_4 sweep of higher speed fluid moving towards the back of the hairpin. The Q_2 event refers to a region of the second quadrant velocity vector, i.e. u<0 and v>0, while the Q_4 event mentioned below refers to the region of the fourth quadrant vectors with u>0 and v<0. The strongest experimental support for the existence of hairpin vortices in the logarithmic layer was originally given by Head and Bandyopadhyay 1981. The instantaneous samples of PIV results show that the attached hairpin packet scenario explains, or at least is consistent with a number of observations of large scale structures explained by Saxton-Fox and Mckeon 2017 and de Silva et al. 2018 in the turbulent boundary layer.

5.5 Temporal-spatial Analysis

Temporal-spatial analysis is utilized for the collected PIV data measured across space and time. This analysis aims to describe the structure phenomenon in a particular location and period of time. Two Reynolds numbers are selected from the PIV dataset (Re_τ= 1667 and 3200) to investigate the propagation of the coherent structure in space and time domain at several wall-normal locations.

Figures (5.8) and (5.9) show the temporal-spatial velocity field at four different wall-normal locations (y/R= 0.1, 0.3, 0.5, 0.7). By inspection, sets of positive and negative stream-wise fluctuation velocities represented in the vertical lines are observed across several short time intervals. These strains give evidence of the large coherent structures presence in turbulent pipe flow. In agreement with the results of fig.(4.7) and (4.8) in chapter(4), large structures at time interval are noticed intensively in the near-wall region at $0.1 \leq y/R \leq 0.5$, where these sets start to disappear by moving towards the pipe axis. Due to the identified wave-length value of such structures $\lambda_x/R > 3$, these energetic structures define to belong to the VLSMs.

On the other side, LSMs can be distinguished in figures (5.8) and (5.9) through their performance, at time interval 1.5, 2, 3, and 4 s for $Re_\tau = 1667$ and 0.25, 1.25, and 2 s for the higher Reynolds number, to show an existence beyond half the pipe radius till y/R =0.7. The obtained results from the temporal-spatial velocity field emphasize the realized conclusion from pre-multiplied contour spectra in the chapter(4), showing that VLSMs are initiated from the near-wall region and vanished at approximately half the pipe axis y/R=0.5, while LSMs manifest to exist along the pipe axis.

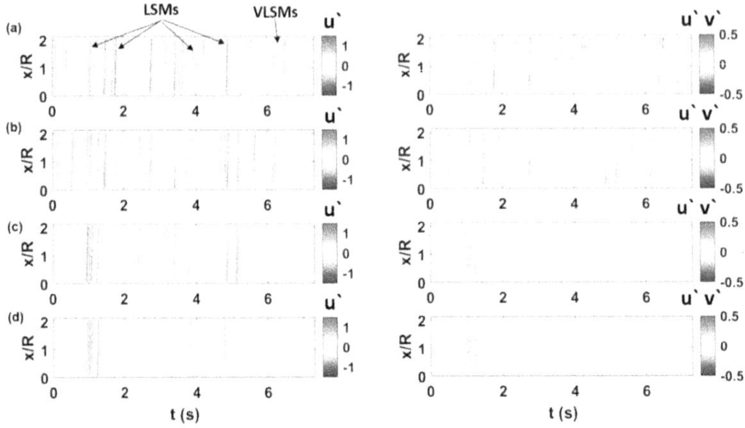

Figure 5.8: Temporal-spatial of the stream-wise fluctuation velocity u' (left) and Reynolds shear stress $u'v'$ (right) for the PIV dataset Re_τ= 1667 at different wall-normal locations. (a) y/R= 0.1, (b) y/R= 0.3, (c) y/R =0.5, and (d) y/R = 0.7.

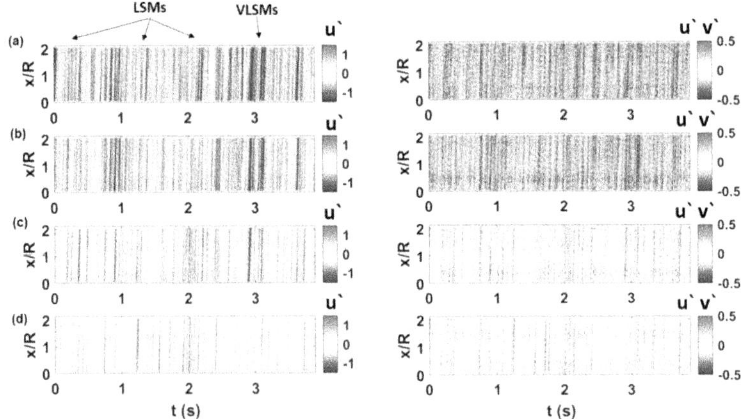

Figure 5.9: Temporal-spatial of the stream-wise fluctuation velocity u' (left) and Reynolds shear stress $u'v'$ (right) for the PIV dataset Re_τ= 3200 at different wall-normal locations. (a) y/R= 0.1, (b) y/R= 0.3, (c) y/R =0.5, and (d) y/R = 0.7.

Reynolds shear stress produced by energy-containing eddies is settled in the near-wall region, as shown in figures (5.8) and (5.9). At shear Reynolds number (Re_τ) =3200, the signature of the Reynolds shear layer can be clearly observed to be dominant at y/R =0.1 and 0.3, whereas Reynolds stress gradient is recognized to be effectively constant in this region.

5.6 Quadrant Analysis

The quadrant analysis of turbulent flow characteristics is a helpful technique to understand the turbulence mechanism of the flow. The instantaneous velocity components (u and v) are used to identify each quadrant representing turbulent events that occur in the flow. These quadrants can be abbreviated as Q_1, Q_2, Q_3, and Q_4 turbulence events.

Sweep Q_4 and eject Q_2 events play a significant role to show and identify the appearance of the energetic structures in wall-bounded flows. Fig.(5.10) presents the energy contribution of turbulent structures at $Re_\tau = 10617$, through the two dominant regions. The figure clarifies the energy concentration at two selected wall-normal locations (y/R =0.1 and 0.8). The near-wall location shows a dominance of Q_2 and Q_4 events compared to the further location y/R= 0.8 due to the intensive presence of coherent structures in the near-wall region. This behaviour supports the conclusion acquired from the temporal-spatial and pre-multiplied spectra analysis that large scale structures exist in the near-wall region than the outer region.

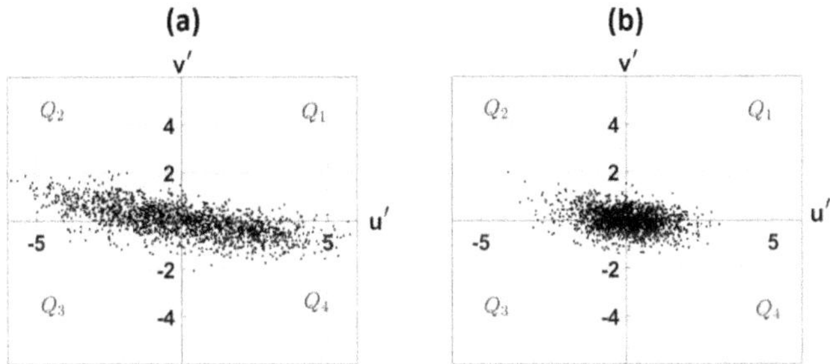

Figure 5.10: Quadrant analysis plots for the fluctuation velocity components (u' and v') at Re_τ=10617. (a) y/R= 0.1, and (b) y/R= 0.8.

Chapter 6: Two-dimensional Spectral Analysis in Pipe Flow

6.1 Introduction

Since Kim and Adrian 1999 discovered the two distinct peaks of large and very large scale motions in the pre-multiplied spectrum of stream-wise fluctuating velocity in pipe flows, coherent structures have received particular attention in the wall-bounded turbulence research community. However, previous authors proposed that the stream-wise alignment of large scale structures represented in a collection of smaller hairpin-shaped vortices are the reason behind very large scale motion formation. Subsequently, their characteristic features and length scale have been verified by considerable experimental and numerical studies in various wall-bounded flow scenarios, such as pipe (Guala et al. 2006; Monty et al. 2007; Bailey and Smits 2010; Hellstrom et al. 2011; Baltzer et al. 2013; Lee and Sung 2013; Lee et al. 2015), close channel (Balakumar and Adrian 2007; Lee et al. 2014, 2015), turbulent boundary layer (Balakumar and Adrian 2007; Lee and Sung 2013). According to the aforementioned studies, the large scale motions are recognized to have a stream-wise scale of approximately 2-3 pipe radii and have been associated with the occurrence of the bulges of turbulence fluid of the edge of the wall layer to induce regions of low stream-wise momentums between the successive hairpins legs.

On the other side, hot-wire rake measurements in internal flows found that VLSMs are defined to be as long as 30 times the channel half-height or pipe radius and reach wavelength > 8-16 R in the outer region of fully developed turbulent pipe flow (Guala et al. 2006; Smits et al. 2011). Furthermore, there is a reasonable agreement on the contribution of VLSMs to the turbulent kinetic energy, but it is still not entirely clear where and how much those structures contribute to the Reynolds shear stress, particularly in pipe flow.

Earlier, it is proved that "large eddies" contribute at least 50% to the turbulent energy and about 80% to the Reynolds stress in the outer layer (Blackwelder and Kovasznay 1972). In addition, Guala et al. 2006; Balakumar and Adrian 2007 used lately spectral analysis of X-probe hot-wire measurements to expound that VLSMs and LSMs produce a significant contribution to turbulent kinetic energy and Reynolds stress in the outer layer. Balakumar and Adrian 2007 found that 40-65% of the kinetic energy and 30-50% of Reynolds shear stress are accounted for in the long modes in the turbulent boundary layer with stream-wise wavelength $\lambda/\delta > 3$, where δ is the boundary layer thickness. A similar investigation in turbulent boundary layer done by Ganapathisubramani et al. 2003 estimates the contribution to Reynolds shear stress from vortex packet structures at moderate Reynolds number in the stream-wise / span-wise planes of the log region.

Although numerous investigations are done on the contribution of coherent structures to total kinetic energy, experimental data on the Reynolds shear stress ($-u'v'$) statistics at high Reynolds numbers are quite rare in turbulent pipe flow, particularly in the time and space domain. For that, the main objective of the present study is to determine how LSMs and VLSMs contribute to total kinetic energy and Reynolds shear stress at a relatively high Reynolds number in turbulent pipe flow in the time and space domain. This will be done via spectral and co-spectral statistical analysis of 1838 captured snapshots of data at two-dimensional planes of high-speed PIV measurements, avoiding onerousness, errors and essential corrections mentioned by other works of literature due to the effects of misalignment and heat impact of the X-probes.(Deshpande et al. 2020)

Throughout this study, the coordinate system x, y refers to the stream-wise (axial) and wall-normal directions (radial), respectively. Corresponding instantaneous stream-wise and wall-normal velocities are represented by u, v respectively, with the corresponding velocity fluctuations given by lower case letters. The subscript "+" refers to normalization by inner scales. For example, one can use $y^+ =$ y u_τ/ν for wall-normal locations and $u^+=$ u/u_τ for stream-wise velocity, where u_τ is the friction velocity and ν is the kinematic viscosity.

6.2 Results

6.2.1 Pre-multiplied Spectral Analysis

For identifying the prevailing scales in the flow field, pre-multiplied wavenumber spectrum is used as in other LSMs and VLSMs studies (e.g. Kim and Adrian 1999; Hutchins and Marusic 2007a; Monty et al. 2009). The pre-multiplied power spectra for the stream-wise velocity components ($k_x \Phi_{uu}(k_x)$) are presented in fig.(6.1) and (6.2).

At lowest Re_τ= 3200, the inner peak footprint of LSMs is apparently observed at $y^+ \geq 60$ in fig.(6.1). However, it is well known that the created main energetic structure (LSMs) in the buffer layer $y^+ < 30$ can coherently align along the stream-wise direction with the transportation roles creating the elongated near-wall streaks with the stream-wise extent of $\sim$ O(1000) (ν/u_τ) (Monty et al. 2009).

Furthermore, the most noticeable feature in fig.(6.1) is the energy contained in the outer peak of VLSMs, marked by black crosses. The peak becomes more distinct when the Reynolds number increases and because of the sufficient inner and outer scale separation as illustrate in the turbulent intensity (TI) profile. The outer peak in pre-multiplied spectra can be observed clearly with its wall-normal position located near the centre of the logarithmic region following the outer peak in TI profile (i.e. $y^+ = 3.9 \sqrt{Re_\tau}$) as highlighted by the vertical black dotted lines), this is similar to what identified in pipe and channel flows (Ng et al. 2011 and Vallikivi et al. 2015).

The outer scaled pre-multiplied spectra in fig.(6.2) show the two characteristics length scales separated by the horizontal line following the terminology of Kim and Adrian 1999. The flow structures associated with shorter wavelength scale (λ) = 3R refers to the LSMs structure, while the larger length scale structure with λ_x/R > 3 indicate to VLSMs. The length scales are determined from the spectral peaks. Two energetic modes in the near-wall and outer layer are demonstrated from fig.(6.2.a) at Re_τ = 3200. Corresponding to the inner and outer peaks of large and very large scale motions, respectively.

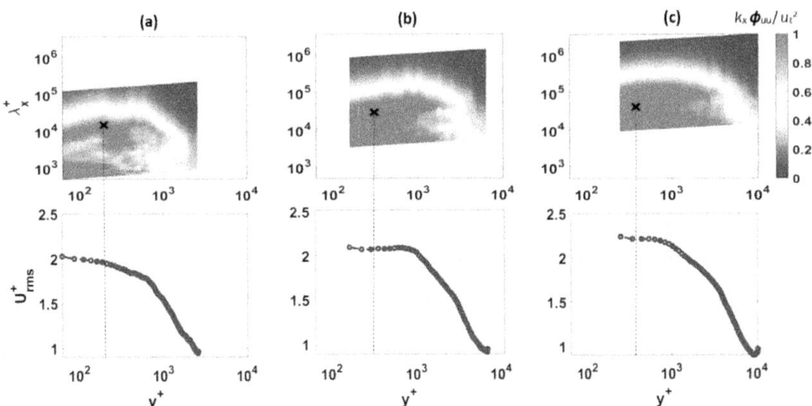

Figure 6.1: Upper row: Contour plots of spectra ($k_x\Phi_{uu}/u_\tau^2$) in inner scale . Bottom row: Turbulent intensity (u^{2+}) profiles. (a) Re_τ = 3200, (b) Re_τ = 6605, and (c) Re_τ= 10617. Symbol (X): location of the outer spectral peak y^+_{osp}.

The length scale of very large structures increases obviously to exist in the outer layer by increasing the Reynolds number, which indicates that VLSMs are the most energetic and dominant structure in the logarithmic region and outer layer. (Smits et al. 2011 and Vallikivi et al. 2015). The separation of LSMs and VLSMs can be observed at higher Reynolds numbers Re_τ)= 6605 and 10617.

Following the widely adopted separation scale value in the literature (Guala et al. 2006; Balakumar and Adrian 2007), λ_x= 3R is used to separate LSMs and VLSMs. Then the contribution of VLSMs to turbulence intensity, or equivalently the fraction of energy carried by VLSMs, can be defined as:

$$\gamma_{uu} = \frac{\int_o^{2\pi/3R} \Phi_{uu}(k_x)dk_x}{\int_o^{\infty} \Phi_{uu}(k_x)dk_x} \tag{6.1}$$

Fig.(6.3) shows the contribution of VLSMs at the three cases in the outer region at y/R $\approx$ 0.2-0.3, where VLSMs dominate γ_{uu} is 0.4-0.55. This result indicates that VLSMs contribute about 55% of total fraction kinetic energy (stream-wise turbulence intensity), which is consistent with the recorded value in Hallol et al. 2020 for one-dimensional stream-wise spectral analysis. Besides, the VLSMs contribution to the fraction kinetic energy in pipe flow is marked to be more significant than in turbulent boundary layer flow (Balakumar and Adrian 2007).

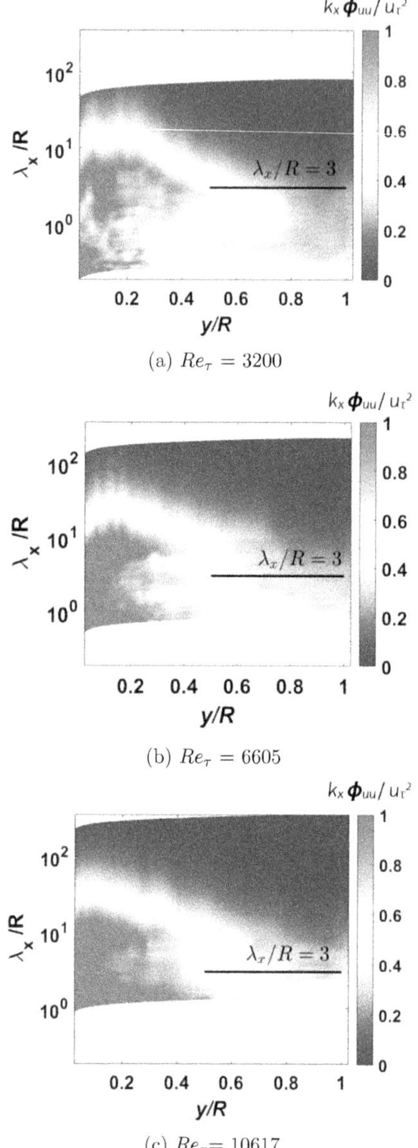

(a) $Re_\tau = 3200$

(b) $Re_\tau = 6605$

(c) $Re_\tau = 10617$

Figure 6.2: Contour plots of spectra ($k_x\Phi_{uu}/u_\tau^2$) in outer scale.

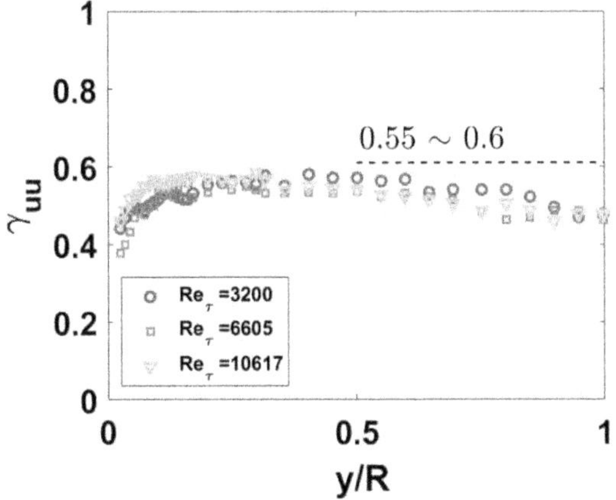

Figure 6.3: Fraction kinetic energy (γ_{uu}) carried out by VLSMs along the wall-normal locations.

6.2.2 Contribution of Reynolds Shear Stress

Pre-multiplied co-spectra of stream-wise and wall-normal velocity $k_x\phi_{uu}(k_x)$ are presented in fig.(6.4) to investigate the contribution of VLSMs to the Reynolds shear stress. The figure shows the inner scaled pre-multiplied co-spectra $k_x\phi_{uv}(k_x)$ for the three measured Reynolds numbers. It is clearly noticed that the coherent structure including VLSMs are more contributing towards the near-wall region for the lowest Reynolds number Re_τ=3200 at $y^+ \leq 2000$, while the major contribution in the outer layer is acquired at a higher Reynolds number and such structures are defined to be corresponding to the VLSMs. This is clarified from the outer scaled co-spectra in fig.(6.5). Whereas the wavelength of VLSMs is about$\lambda_x \leq$ 3R while lengthscale of VLSMs is greater than 3R based on the peak amplitude of the pre-multiplied spectra at large and very large scale motions wavelengths. This performance supports via the reference line separating both structures. The figure highlights the situation in the outer layer that presents approximately similar significant behaviour of energy contribution as the power pre-multiplied spectra of stream-wise fluctuation presented in fig.(6.4).

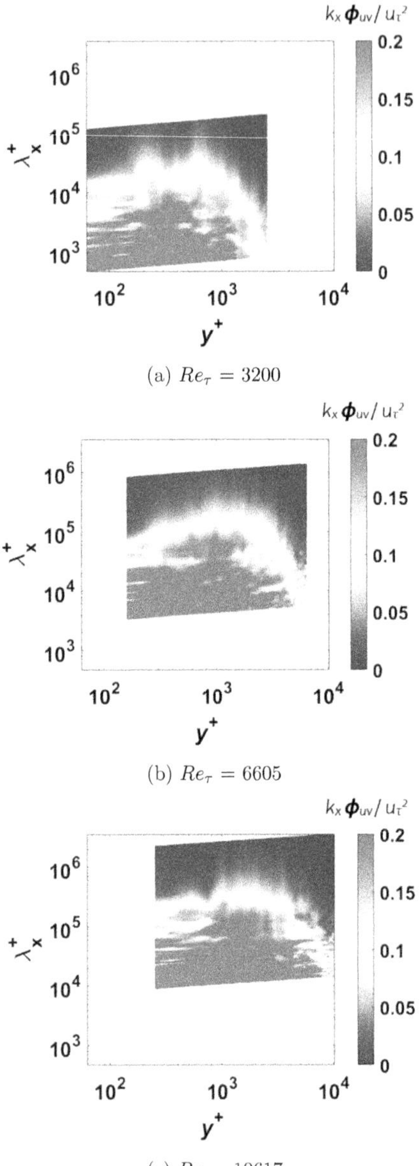

(a) $Re_\tau = 3200$

(b) $Re_\tau = 6605$

(c) $Re_\tau = 10617$

Figure 6.4: Contour plots of co-spectra ($k_x \Phi_{uv}/u_\tau^2$) in inner scale.

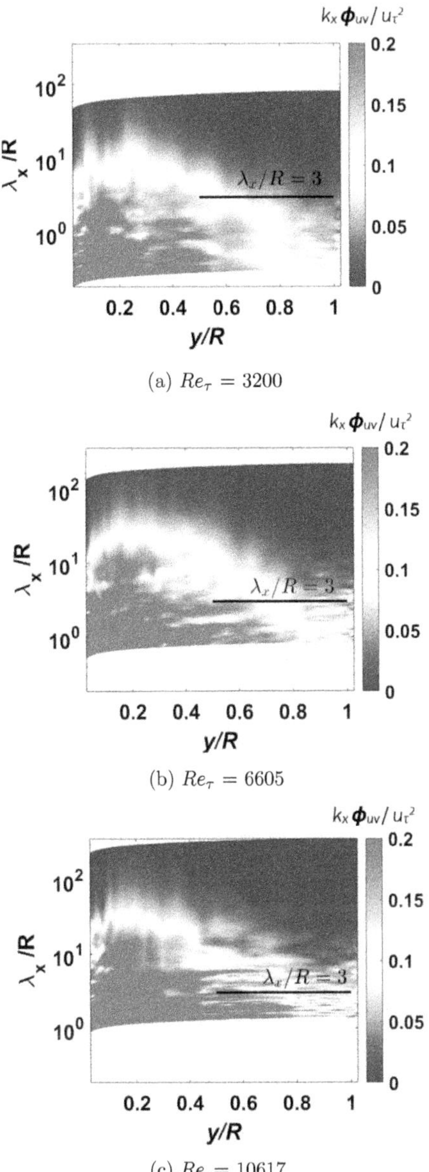

(a) $Re_\tau = 3200$

(b) $Re_\tau = 6605$

(c) Re_τ= 10617

Figure 6.5: Contour plots of co-spectra $(k_x \Phi_{uv}/u_\tau^2)$ in outer scale.

By following the definition of γ_{uu}. The fraction of Reynolds shear stress carried by the VLSMs, γ_{uv} can be obtained in identical ways and the wall-normal variations of γ_{uu} are presented in fig.(6.6).

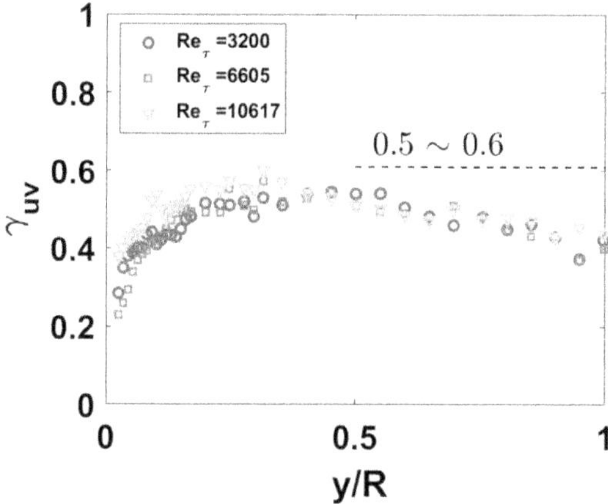

Figure 6.6: Fraction Reynolds shear stress (γ_{uv}) carried out by VLSMs along the wall-normal locations.

Here, γ_{uv} increases monotonically between 0 < y/R < 0.5, at y/R=0.3. γ_{uv} reaches approximately 0.45 ~ 0.6. Then for y/R ≥ 0.5, γ_{uv} decreases slightly to obtain constant Reynolds shear stress gradient in this region towards the pipe axis. It is concluded over all of the y locations and Reynolds number that VLSMs contribute with 45% to 60% to the Reynolds shear stress in the logarithmic region and outer layer, which is larger than what has been demonstrated in the turbulent boundary layer.

The spectral analysis results agree with the previous flow scenarios that VLSMs are both energetic and stress across the entire outer region. In addition, it indicates that as streaky motions are the major contributors in the near-wall region to both stream-wise turbulent intensity and Reynolds shear stress, VLSMs are provided to be the major contributors in the logarithmic region and outer layer to turbulent kinetic energy and Reynolds shear stress.

6.3 Conclusion

Experimental study of very large scale motions in CoLaPipe test facility has been quantified via turbulence statistics based on two-dimensional time-resolved particle image velocimetry experimental data. The experiment is carried out at relatively high friction Reynolds numbers Re_τ =3200, 6605, and 10617. Based on the comparison of the present datasets with the reported experimental and numerical works of literature, the contour of stream-wise pre-multiplied spectra and Reynolds shear stress in u and v velocity components indicate good agreement.

Pre-multiplied power spectra and co-spectra analysis combined with the scale separation are used to determine the contribution of VLSMs to the turbulent statistical quantities, including total kinetic energy and Reynolds shear stress. Major findings are summarized as follow:

- The obtained pre-multiplied spectra exhibit the presence of VLSMs and the footprint of LSMs where the spectral content implies mainly the relationship between VLSMs and turbulent kinetic energy distribution phenomena. The VLSMs contribute about 55% to total kinetic energy of equivalent association of small and large scale structures.
- The Reynolds number effects are assessed and found to have minimal influence on VLSMs length scale, though the amplitudes of the pre-multiplied spectra and co-spectra demonstrate the dependence of Reynolds number. This consistent with the turbulent boundary layer explanation (Balakumar and Adrian 2007; Hutchins and Marusic 2007a)
- The contribution of VLSMs is mainly apparent in the two-dimensional spectra (co-spectra uv), the VLSMs provide about 60% to Reynolds shear stress in the logarithmic region. This result shows the evidence of VLSMs and how they play an important role in the outer layer. It is also proved that VLSMs are the major contribution to total kinetic energy and Reynolds shear stress in the logarithmic and outer layer.
- Protracted flow field is required in future work to obtain sufficient spatial resolution over a longer stream-wise extent. However, large coherent structures (i.e. very

large scale motions) are recognized to commonly persist with a length greater than 20 R (Adrian and Marusic 2012).

Chapter 7: Proper Orthogonal Decomposition Analysis (POD)

7.1 Introduction

Proper orthogonal decomposition (POD) is a technique that is used significantly in recent years for solving and analysing the characteristics of flow fields. This method facilitates the analysis of large data sets by reducing the order distribution process. POD is known as Karhnnen-Loéve decomposition, principal components analysis, singular system analysis, and singular value decomposition (Holmes et al. 1997). Lumley 1967 was the first to relate the POD method to turbulent flow. He pioneered the use of POD to study the internal structures in inhomogeneous turbulence (Yang et al. 2019). Since that time, the method of POD has served as an effective tool to extract dominant coherent structures through structural analysis of generic turbulent flow based on energy contribution (Yin et al. 2019). Meanwhile, the POD technique was applied together with the PIV technique (PIV-based POD) to successfully resolve the Large-scale coherent structures of the turbulent boundary layer (Jin and Ma 2018, Li et al. 2018, Güemes et al. 2019) and turbulent pipe flows (Duggleby et al. 2007, Hellström and Smits 2017, Antoranz et al. 2018).

The snapshot POD method proposed by Sirvich et al. 1987 is used to visualize the very Large-scale motions (VLSMs) in fully developed turbulent pipe flow. Hellström et al. 2011 suggested a possible connection between the origin of the VLSMs and linear stability analysis. Such structures can be reconstructed using a small number of POD modes. Only the four most energetic modes is needed to recreate the meandering structures that appear to be much longer than any of its constituent modes. Although the origin of VLSMs is not yet sufficiently clear, Guala et al. 2006 demonstrated that much of the energy and Reynolds shear stress at attributed to the lowest-order POD modes establish in the VLSMs. They proved that spectral analysis of the Y-derivatives of Reynolds shear stress presents the same smaller motions, including the LSMs and main turbulent motions contribute the same amount as VLSMs to the net Reynolds shear force, (d-$u'v'$/dy). Liu et al. 2001 noticed that large scale structures acquired from projecting sample velocity fields onto the dominant modes resemble the signature of a hairpin vortex (Wu 2014). The strong contribution of the hairpin vortex to the dominant POD modes in the near-wall region is observed in both wall-normal and stream-wise directions (Baltzer and Adrian 2011). These observations are consistent with Hellström et al. 2011 results which demonstrated that the first ten modes capture well the large scale structures.

Large scale structures are identified through the second (or fourth) quadrant vectors of Q_2 (ejection) or Q4 (sweep) events, respectively. Discetti et al. 2019, recently applied an extended POD-based dynamic estimation to define and measure the duration of 'global' low momentum (ejection) structure and high momentum (sweep) structure events based on the most energetic modes and the reconstructed time coefficient. These results suggest that the first mode is largely dominant with an independent Reynolds number. Otherwise, low-order modes are recognized as the most energetic modes that carry significant contributors to the shear stress. However, the first ten modes contribute about 15% of the turbulent kinetic energy and 43% of the integrated shear stress (Hellström et al. 2015). In contrast, Wu 2008 and Wu et al. 2010 reveal that the first POD mode occupies the most amount of kinetic energy in both channel flow and boundary layers.

A new application of the POD method was devised by Wu 2014 to identify the contribution of instantaneous turbulent structures to the first and second POD modes in the turbulent boundary layer. Based on his method, the present work aims to study how the dominant large-scale structures contributing significantly to the first POD affect the turbulence single and two points statistics in turbulent pipe flow.

7.2 Proper Orthogonal Decomposition

Applications of the snapshot POD method have higher computational efficiency to decompose the random vector field turbulent fluid motions into a set of deterministic functions that each mode contains a portion of the total fluctuating kinetic energy in the flow (Kellnerová et al. 2012). This method is applied in the present study to mode-decompose the PIV measured velocity vector fields of turbulent pipe flow. The

POD analysis using the velocity fluctuations as the targeted variable (Sirovich et al. 1987). Herein the fluctuating velocity components are indicated with u' and v' for x and y directions respectively. The mean velocity field is subtracted from transient flow fields (snapshots) to generate the fluctuating velocity fields. In the two-dimensional (2D) POD, any instantaneous velocity fluctuation in the stream-wise (x) / wall-normal (y) plane, u'(x,t) can be decomposed into the form of :

$$u'(x,t) = \sum_{n=1}^{N} a_n(t)\Phi_n(x) \tag{7.1}$$

Where x is the coordinate of stream-wise direction and t represents either time-independent or time-dependent snapshots of the PIV measured velocity fields. The eigenfunction of POD modes indicated to the deterministic spatial POD modes Φ_n (x), n^{th} the number of modes while a_n(t) is the POD temporal coefficient which is calculated by solving the eigenvalue problem with positive definite Hermitian Kernel of the form

$$\lambda_n a_n(t) = \int_T \left(\int_\Omega u'(x,t)u(x,t)dx \right) a_n(t')dt' \tag{7.2}$$

For an ensemble of measured velocity fields, the spatial domain Φ could be the whole or part of the field of view (FOV), while the time domain (T) refers to the ensemble or the collection of samples of the velocity fields. The integration in equation (2) is shown to be over both domains.

According to Wu et al. 2017, the eigenvalues λ_n are exhibited to be real and positive to form a decreasing and convergent series. In addition, they reported that the larger value of $a_n(t)^2$ the more contribution to λ_n. This is concluded through the computed POD modes equation

$$\Phi_n(x) = \frac{\int_T u'(x)a_n(t)dt}{\int_T [a_n(t)]^2 dt} \tag{7.3}$$

, and the eigenvalues of the normalized orthonormal POD modes, i.e.;

$$\lambda_n = \int_T [a_n(t)]^2 dt \tag{7.4}$$

As known turbulent kinetic energy (TKE) is equal to half of the summation of the eigenvalues, i.e.;

$$TKE = \frac{1}{2}\sum_{n=1}^{N} \lambda_n \tag{7.5}$$

The important temporal coefficient of $|\ a_n(t)|$ of POD deduced to contain equivalent important information on the major contributing flow structures to POD modes (Wu et al. 2014). This method can be utilized to determine the instantaneous fluctuated flow structures that contribute significantly to the first low order dominant POD (Yang et al. 2019). The contribution from the first mode is proposed to be largely dominant compared to the second mode (Discetti et al. 2019, Zhang et al. 2014 and Chen et al. 2012). For that, the current study concerns with the influence of turbulent fluctuation in the first mode that explained to be the most energetic POD mode.

7.3 POD Analysis

POD method has decomposed the datasets of 1838 instantaneous fluctuation velocity fields. Snapshot POD is performed in the two components fluctuating velocity data. The turbulent kinetic energy fraction contribution of the first 20 modes is given in fig.(7.1). It is noticed that the first POD mode accounts for more than 17% of the total energy while the first two modes in sum are about 23%. Fig.(7.1) shows that the most turbulent kinetic energy (TKE) is intensively contained within the first and second-order mode, and the normalized TKE in the higher modes decreases dramatically as the eigenmode increases. Similar results of energy contribution of POD modes deduced in other studies, Xu et al. 2020 and Wu 2014. In addition, Discetti et al. 2019 mentioned that large scale motions (corresponding to the first POD mode) carry more than 40% of TKE in the turbulent boundary layer if the modes from one to three are taken into consideration. POD eigenspectra (λ / $\sum_{n=0}^{N} \lambda_i$) and cumulative energy ($\sum_{n=0}^{n} \lambda_i$ / $\sum_{n=0}^{N} \lambda_i$) for the two measured Reynolds numbers $Re_\tau = 6605$ and 10617 are illustrated in fig.(7.1).

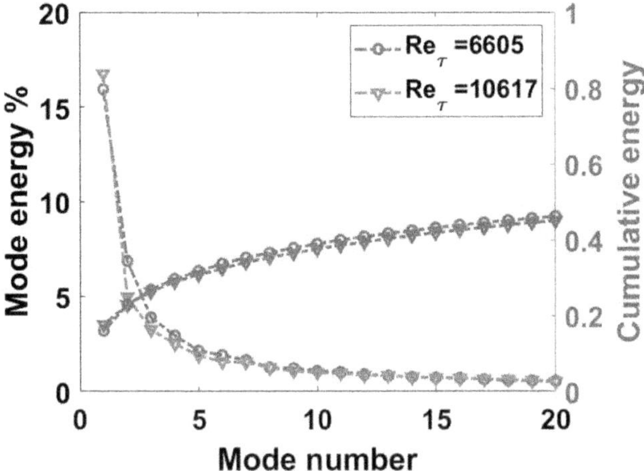

Figure 7.1: Kinetic energy content and cumulative energy of the first 20 POD modes.

The in-plane TKE is denoted to the first mode with a magnitude of about 17% as presented in fig.(7.1), no Reynolds number dependence is observed. Therefore only Re_τ = 10617 is presented for brevity in the current study.

Consequently, the contours of the first three POD modes of the stream-wise component are shown in fig.(7.2). The first POD contour in fig.(7.2.a) depicts an extension of a large scale fourth-quadrant (Q_4) is observed for the whole field of view (FOV). However, the dominating sweep event Q_4 exits intensively in the region y/R$\leq$ 0.65. The dominating event in the first POD mode is either sweep or ejection depending on the multiplied positive or negative values of its time coefficient (a_1) (Vila et al. 2017, Lozano-Duràn and Jiménez 2014, and Munir et al. 2015).

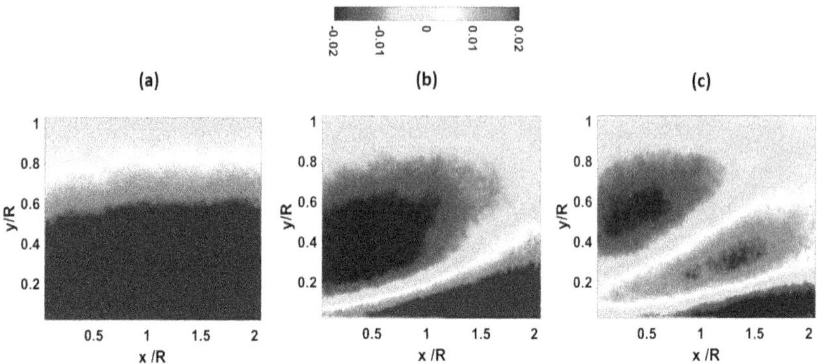

Figure 7.2: POD spatial modes of the velocity fields (a) First mode, (b) Second mode, and (c) Third mode at $Re_\tau = 10617$.

The Reynolds ejection and sweep events with different signs of these velocity fluctuations (Q2:-u', +v'; Q4: +u', -v') produce the major association to the Reynolds shear stress and the turbulent kinetic energy production in channel and pipe flows (Tang et al. 2012; Yang et al. 2012; and Wallace 2016). On the other side, a shear layer in the second mode across the whole FOV is exhibited in fig.(7.2.b), the layer extends in the stream-wise direction to induce an inclination angle of 20° from the wall. This layer which separates the ejection event (Q_2) from the sweep event (Q_4), indicates an interaction of Reynolds stress events. The strong sweep/ejection events in the first mode ($\Phi_n = 1$), as well as the shear layer in ($\Phi_n = 2$) for the adjacent boundary layer, are similarly observed in the classical turbulent layer (Brown and Thomas 1977, Head and Bandyopadhyay 1981, Österlund and Johansson 1999, and Yang et al. 2019) and observed in turbulent pipe flow by Discetti et al. 2019. The third and fourth POD modes share approximately a similar amount of energy, for that this study presents only the third mode. From mode three, the thickness of ejections / sweeps can be determined. It is observed from fig.(7.2.c) that the positive time coefficient of mode three makes the sweep thinner, while the ejection starts to shrink in the wall-normal direction $0.3 \leq y/R \leq 0.85$ and extends in the stream-wise direction close to the wall. This phenomenon agrees with Discetti et al. 2019 observations in turbulent pipe flow.

Positive and negative values in the first POD coefficient (a_1) are presented through the scatter plot and histogram in fig.(7.3.a). The coefficient a_1 is normalized by its root mean square (RMS) value σ_{a_1} . The scatter plot clarifies that a small amount of velocity field with positive and negative values a_1 is beyond twice its normalized value. Although the fluctuating velocity fields $\geq\pm$ 2 σ_{a_1} are few, it is coefficient larger values a_1 carry a significant weight that increases the contribution to the eigenvalue λ_n in the first POD mode (i.e. to a part of TKE).

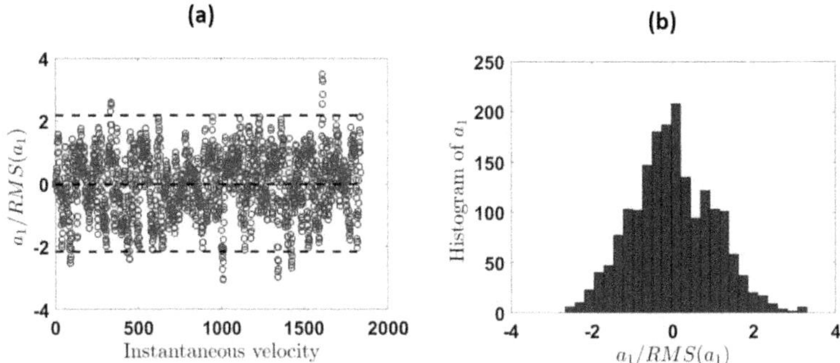

Figure 7.3: (a) Scatter plot of the first POD time coefficient (a_1) normalized by its RMS value for Re_τ=10617, and (b) Histogram of normalized a_1.

This conclusion was proved by Wu 2014 and discussed in an equation (7.4). Whereas the structures or motions in the instantaneous fluctuating velocity field, the larger positive (negative) value of a_1, the more energetic large scale structures, or the stronger Q_2 (Q_4) events associated with the first mode. Approximately symmetric distribution values of positive and negative coefficient a_1 are noticed through the histogram, showed in fig.(7.3.b). The similarity in the histogram indicates the same contribution from Reynolds ejection Q_2 and sweep Q_4 events to the first POD mode.

Qualitatively similar Q_2 and Q_4 events in the first POD mode which show in fig.(7.2) are presented in a separated instantaneous fluctuating velocity fields as presented in figures (7.4) and (7.5). The two velocity fields contain a relatively large positive coefficient a_1 = +2 σ_{a_1} and negative coefficient a_1 = -2 σ_{a_1} separately for the first POD mode. In fig.(7.4), the main feature of the field of positive coefficient a_1 is the large-scale sweep event that attains to 0.5 $\leq$ y/R $\leq$ 1 and extends to 1.6R in the stream-wise direction. As well, a strain of outer layer vortex packets are observed at 0.3 $\leq$ y/R $\leq$ 0.6 beyond the near-wall region $0.6 \leq$ x/R ≤ 1.6. Herein, the Q_2 and Q_4 vectors interact with each other to form small vortex packets close to the wall $0.4 \leq$ x/R ≤ 1.3 at $0.1 \leq$ y/R $\leq$ 0.16. Those vortex cores are the result of individual hairpins within each packet in the near-wall region. Such structures are similar to the hierarchy of hairpin vortex packets explained by Adrian 2007 in the turbulent boundary layer.

Otherwise, fig.(7.5) illustrates an instantaneous fluctuation velocity field with a negative coefficient of a_1 =-2 σ_{a_1}. It appears that the Q_2 (ejection) event is decomposed intensively at the right upper region and concentrated very close to the wall to form a small number of vortices. Furthermore, Q_4 vectors are distinguished in the middle region of the FOV. It is not clear if the presence of the upper left small vortex is a result of the hairpin signature of the Q_2 event due to the variation of low and high momentum or the velocity convection. 3D PIV investigation is recommended to give more clarification for the observed phenomena.

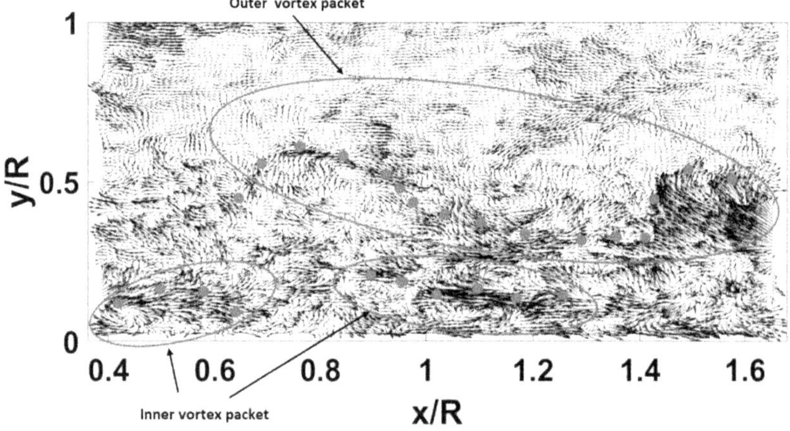

Figure 7.4: An instantaneous fluctuating velocity field with a large positive POD coefficient of $a_1 = 2.17\sigma_{a1}$ for the first POD mode.

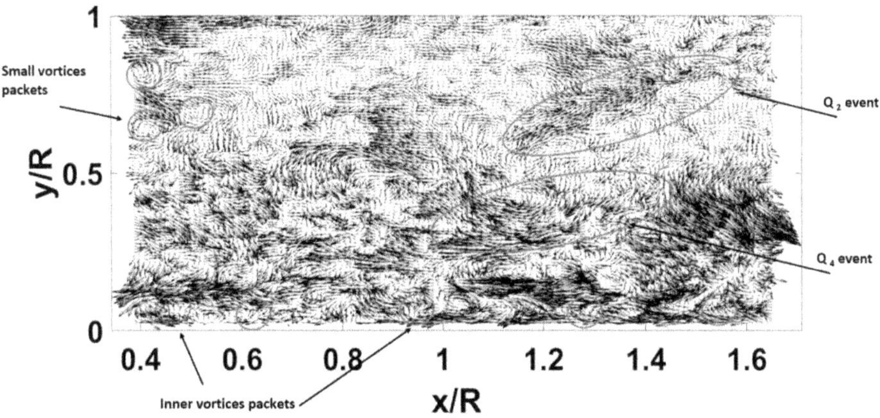

Figure 7.5: An instantaneous fluctuating velocity field with a large negative POD coefficient of a_1 = -2.17σ_{a1} for the first POD mode.

7.3.1 Contributions to Reynolds Stresses (-uv^+, u^{2+}, and v^{2+})

For more consideration, the contribution of large scale structures to the first POD mode is investigated in the current study by following Wu 2014 method that is implemented in turbulent boundary layer flow. This method depends on obtaining the new ensemble of fluctuating velocity field after removing the small number of the field that exceeds $\pm 2\ \sigma_{a_1}$ from the original ensemble of 1838 fields. Whereas those removing fluctuating velocity fields are the most dominant contributors to the first POD Wu 2014.

The affection of those high energetic fields of the turbulent Reynolds stress in the stream-wise and wall-normal directions and the Reynolds shear stress is identified by comparing the original ensemble fields with the newly obtained ensembles without the fluctuating fields that carry high energy. Those removed portion fields are about 8.5% from the original ensemble of 1838 fields. Fig.(7.6.a), shows that the data of the complete velocity field ensembles obtain a higher magnitude (about 6%) of Reynolds stress in the stream-wise direction comparing to the data of the new ensemble fields. This higher magnitude of Reynolds stress can be clarified well at y/R $\leq$0.4. In the radial or wall-normal direction, as indicates in fig.(7.6.b), It is observed that there is no noticeable difference in the normalized Reynolds stress for both different ensembles (i.e., with and without extra fluctuating fields> $\pm$ 2 σ_{a_1}).

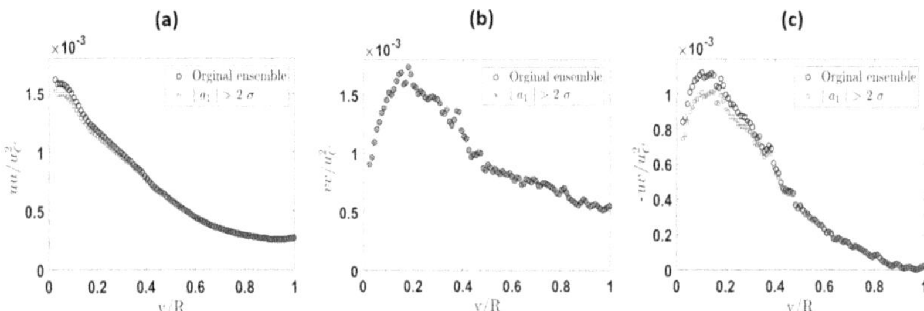

Figure 7.6: Comparisons of Reynolds stresses between the original ensemble of the fluctuating velocity fields and the ensemble without those velocity fields whose $\mid a_1 \mid > 2\sigma_{a1}$. (a) Stream-wise Reynolds stress (u^2); (b) Wall-normal Reynolds stress (v^2); and (c) Reynolds shear stress ($-uv$).

The Results showed that large scale structures interpreted in the fluctuating velocity fields contribute significantly to the large amount of TKE to stream-wise Reynolds stress. Otherwise, the comparison of Reynolds shear stress in fig.(7.6.c) presents a salient deviation between the original ensemble field and the removed portion fluctuating velocity field. This difference can be identified well at y/R $\leq$ 0.5 with approximately 9.3% that is likely to be heavily obtained from u velocity component.

Wu 2017 explained that the observation of Reynolds shear stress in the turbulent boundary layer is expected because the removed large scale structure from the original ensemble belong to large scale Q_2 and Q_4 events that are essential in Reynolds shear stress contributors in the outer layer. This conclusion consistent with Guala et al. 2006, who suggested that relatively small structures, approximately 1R in the near-wall, are responsible for the majority of the Reynolds shear stress, and larger structures with scales up to 5R in the logarithmic layer are responsible for the majority of the Reynolds shear stress. Adrian et al. 2000 demonstrated that this is the flow region where Q_2 ejections of the hairpin dominate over Q_4 events. This can be realized well in fig.(7.4) and (7.5). Furthermore, Liu et al. 2001 observed that the large scale POD mode containing half of the TKE and contain two-third to three-quarter of the Reynolds shear stress in the outer region of the channel flows.

7.3.2 Q_2 and Q_4 Events Contribution in Reynolds Stresses

More investigation of large scale structures contribution is followed by computing two new ensembles from the original POD time coefficient, $| a_1 | > 1.5\ \sigma_{a_1}$ and $| a_1 | > 1\ \sigma_{a_1}$ with 30% and 53% of flow fields respectively. This can be presented clearly in fig.(7.7). The figure shows the comparison of stream-wise Reynolds stress (u^2) and Reynolds shear stress (-uv) at different ensemble of instantaneous velocity fields whose absolute a_1 values are greater than 2 σ_{a_1}, 1.5 σ_{a_1} and 1 σ_{a_1}.

As observed from fig.(7.6.b), contribution to wall-normal Reynolds stress v^2 is almost non-existent. Therefore it's not presented in fig.(7.7). It indicates that the magnitude of u^2 of the lowest ensemble is reducing by increasing the number of removable time coefficient ensemble to deviate with 24% of energy contribution. This result is expected due to the large magnitude of the removed fluctuating velocity field of $| a_1 | > 1\sigma_{a_1}$, which is represented to be a large contribution to turbulent kinetic energy. On the other side, the contribution to Reynolds shear stress in fig.(7.7.b) introduces an obvious higher percentage by decreasing the number of the fluctuating velocity fields to acquire 9.3% for higher ensemble and 20 % for the lower ensemble.

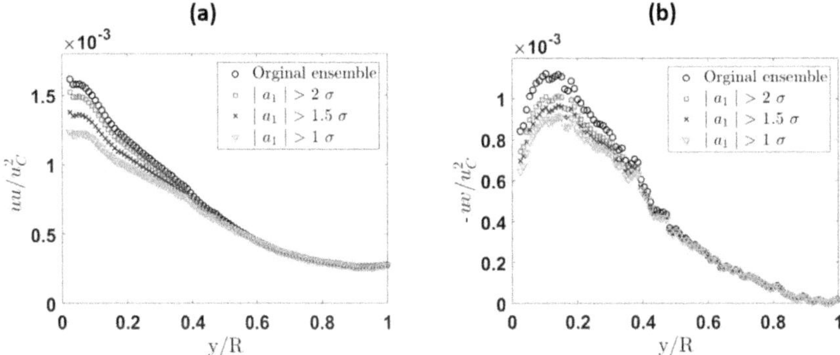

Figure 7.7: Comparisons of Reynolds stresses between the original ensemble of the fluctuating velocity fields and the ensemble without those velocity fields. (a) Stream-wise Reynolds stress (u^2); and (b) Reynolds shear stress ($-uv$).

It is noticed through both figures (7.7.a) and (7.7.b) that the influence of turbulence structures is distinguished intensively toward the wall region at y/R = 0-0.5, and the affection starts to slightly contract further from the wall representing in a coincide for all ensembles at y/R > 0.45. The behaviour of Reynolds shear stress is explained to be depending on the consistent presence of turbulence large-scale structures in the near-wall region that plays an important role in the contribution to Reynolds shear stress in this region. The behaviour of -uv agrees with Wu 2014 observation in turbulent boundary layer except that ensembles contractions are extended to be at wall-normal location y/R > 0.6 instead of 0.45 as exhibited in the present study of turbulent pipe flow. This behaviour is normally expected due to different characterization between the turbulent boundary layer and turbulent pipe flow.

7.4 Conclusion

An approach of a new POD method is performed in turbulent pipe flow to study the contribution of large scale structures to turbulent statistics in the first POD mode. This method has diverse by Wu 2014 and implemented previously in a turbulent boundary layer. The symmetric pattern of Gaussian distribution for positive and negative values of POD temporal coefficient a_1 indicates that the first POD mode carries the same contribution from Reynolds Q_2 and Q_4 events correspondingly with values larger than twice of their normalized RMS values, i.e. $\mid a_1 \mid \geq 2\ \sigma_{a_1}$.

Based on this symmetry the instantaneous velocity fields of analysed 1838 snapshots are computed into two separated velocity vector fields $\mid a_1 \mid > 2\sigma_{a_1}$ and $\mid a_1 \mid > -2\sigma_{a_1}$. Consequently, it is realized that the most energetic structures are presented as many large scale Q_2 or Q_4 events, and small vortex are almost prograde at the near wall. Besides, hairpin vortex packet and large scale Q_4 event are identified as the major structures and the most important contributors to the first POD mode. This demonstration is consistent with Wu 2014 and Discetti et al. 2019. The contribution of large scale structures to stream-wise Reynolds stress u^2, wall-normal Reynolds stress v^2 and Reynolds shear stress -uv is identified by computing new ensembles of removed fluctuating velocity fields (i.e $\mid a_1 \mid > 2\ \sigma_{a_1}$, $\mid a_1 \mid > 1.5\ \sigma_{a_1}$ and $\mid a_1 \mid > 1\ \sigma_{a_1}$) and comparing it with the original velocity field ensemble.

It is observed in the stream-wise Reynolds stress that u^2 value is reduced by increasing the number of removable time coefficient ensembles a_1 to attain 6% for the highest ensemble $\mid a_1 \mid > 2\ \sigma_{a_1}$) and 24% for the lowest ensemble ($\mid a_1 \mid > 1\sigma_{a_1}$). A significant coincide is noticed for all ensembles at y/R $\leq$ 0.5. No difference in wall-normal Reynolds stress v^2 is observed. Furthermore, the contribution of Q_2 and Q_4 events to Reynolds shear stress -uv is realised to be more influenced toward the wall region at y/R = 0-0.5 to acquire 9.3 % of the contribution for the highest ensemble ($| a_1 | > 2\ \sigma_{a_1}$) and 20 % for the lowest ensemble ($| a_1 | > 1\ \sigma_{a_1}$).

Chapter 8: Conclusions and Final Remarks

CoLaPipe test facility was used to generate a wide range of Reynolds numbers up to 1.1×10^6, to study the coherent turbulent structures (large and very large scale motions) in fully developed turbulent pipe flow. The Reynolds number range for the recent experiment is 8×10^4 - 1×10^6. The results indicate the following:

8.1 Comparison Between In-situ and Ex-situ Calibration Methods

HWA measurements precision is dependent on the usable calibration method. For that, Comparing calibration methods with each other will assist in explaining the efficiency of each method, as well it proposes which method is convenient for precise hot-wire final results.

- Curves clarified that measuring data corresponding to In-situ calibration; exhibits a satisfactory behaviour comparing to Ex-situ calibration data. Although kurtosis profiles for both methods show a good collapse with DNS data of Penga et al. 2018 at selected Reynolds numbers.

- In-situ method presents a good agreement with DNS data in skewness than Ex-situ that accomplishes a deviation range of 10-15 $\pm 0.5\%$. Nevertheless, the behaviour in skewness demonstrates that the deviation of measuring data regarding Ex-situ calibration increases with increasing Reynolds number.

- Ex-situ calibration method shows a distinguishable effect on hot-wire measurements accuracy observed by the deviation consequence range. This deviation is identified through the detection of the outer peak in the contour of pre-multiplied spectra. The location of the peaks is found to be different at both methods.

8.2 One-dimensional Spectral Analysis

- The results of the spectral analysis of stream-wise velocity done using Hot-wire anemometer at four different Reynolds numbers Re_τ = 2156, 6556, 13132 and 19000 revealed similar recognized characteristic features of large scale motions (LSMs) in fully developed turbulent pipe flow. LSMs shows approximately constant performance along the pipe axis with maximum wavelength value of λ_x/R =3-3.5.

- Otherwise, the very large scale motions (VLSMs) present a slowly increase in its wavelength value to attain 19 R for maximum Reynolds number range Re_τ = 19000 in CoLaPipe facility and persist in the outer layer till half pipe radius y/R= 0.5.

- Throughout the power spectra, the k_x^{-1} slop is detected in the two experimental cases at low wavenumber region $k_x < 1$, at high Reynolds numbers Re_τ = 6556 and 19000 the k_x^{-1} region is only evident over a very limited spatial location y/R = 0.08 due to the physical extent of wall layer at high Reynolds number. This inference is consistent with Nickels et al. 2005 investigations.

- Furthermore, outer peaks in pre-multiplied spectra are monitored obviously at radial locations close to the wall direction and roughly distinguishable towards pipe axis with increasing Reynolds number. At the same time, the inner peaks appeared to be converge in the inner scale wavenumber spectra.

- The location of the outer spectral peak (y^+_{osp}) is detected by using Mathis et al. 2009 estimation to identify at normalized wall-normal location $y^+ \approx 317$ for Re_τ = 6556 in the contour plots and its location increases with increasing Reynolds number. The y^+_{osp} in pre-multiplied spectral contours are observed to follow the outer spectral peak in turbulent intensity, which has been demonstrated earlier by Vallikivi et al. 2015.

- The outer peaks are obtained at wavelength $\lambda_x > 3$R to present the largest energy content outside the viscous wall region. The cumulative energy fraction estimation is identified based on the wavelength value of very large and large scale structures corresponding to the low and high wavenumber peaks in the one-dimensional pre-multiplied power spectra, respectively. Consequently, it realized that VLSMs with wavelength value greater than 3R carry 55% of the energy of the stream-wise velocity component in the outer layer y/R = 0.28 (the most energetic region).

8.3 Two-dimensional Spectral Analysis

Experimental study of very large scale motions in CoLaPipe test facility has been quantified via turbulence statistics based on two-dimensional time-resolved particle image velocimetry experimental data. The experiment is carried out at relatively high friction Reynolds numbers Re_τ =3200, 6605, and 10617. Pre-multiplied power spectral and co-spectral analysis combined with the scale separation are used to determine the contribution of VLSMs to the turbulent statistical quantities, including total kinetic energy and Reynolds shear stress. Major findings are summarized as follow:

- The visualized instantaneous snapshots provide a distribution of vorticity in the logarithmic region characterized by spatially coherent packets of hairpin vortices. These hairpin groups forming large and very large structures are noticed to be inclined at a characteristic angle of 15-20° to the wall and agree with Liu et al. 1991; Saxton-Fox and Mckweon 2017 observations in flow near-wall shear layers.

- The obtained pre-multiplied spectra exhibit the presence of VLSMs and the footprint of LSMs where the spectral content implies mainly the relationship between VLSMs and turbulent kinetic energy distribution phenomena. The energy-containing structures corresponding to VLSMs carry about 55% to total kinetic energy of equivalent association of small and large scale structures.

- The results agree with the achieved results of one-dimensional spectral analysis in chapter(4).This agreement can be clearly shown through the comparison of the pre-multiplied spectra of stream-wise velocity fluctuation for the two implemented measurement technique, HWA and PIV datasets. Although the small and large structures are not well resolved in the near-wall region using the PIV system representing in the partial disappearance of the inner peak as shown in fig.(8.1). The selected pre-multiplied shows a significant agreement with HWA datasets. In addition, the detection of the outer spectral peak in fig.(8.2) identified the VLSMs that carry the same energy content at a similar selected Reynolds number (Re_τ =6556). The obtained results support the accuracy of both measurement techniques.

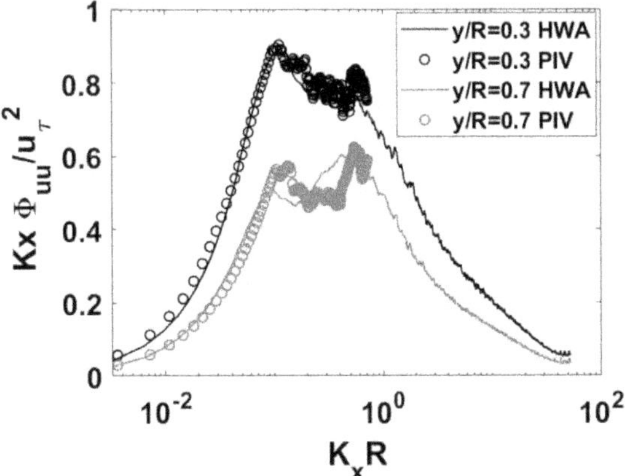

Figure 8.1: Comparsion between the pre-multiplied spectra of stream-wise velocity fluctuation obtained using HWA and PIV datasets at the same Reynolds number Re_τ =6556.

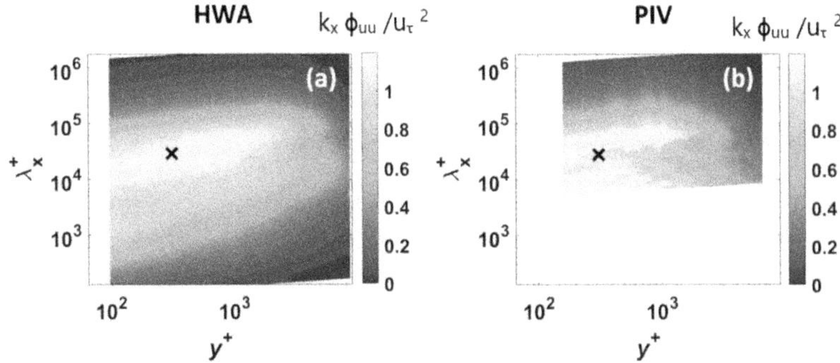

Figure 8.2: Pre-multiplied contour spectra (inner scaled) of stream-wise velocity fluctuation for the two implemented measurement technique, HWA and PIV datasets at Re_τ =6556.

- Temporal-spatial analysis of time-resolved PIV datasets at Re_τ= 1667 and 3200 shows a set of positive and negative stream-wise fluctuation velocities represented in the vertical lines across several short time intervals. These strains give evidence of the large coherent structures presence in turbulent pipe flow. The results support the conclusion of spectral analysis, which explain that VLSMs are initiated from the near-wall region and vanished at half the pipe axis y/R=0.5, while LSMs manifest to exist along the pipe axis.

- The Reynolds number effects are assessed and found to have minimal influence on VLSMs length scale, though the amplitudes of the pre-multiplied spectra and co-spectra demonstrate the dependence of Reynolds number. This consistent with the turbulent boundary layer explanation (Balakumar and Adrian 2007; Hutchins and Marusic 2007a)

- The contribution of VLSMs is mainly apparent in the two-dimensional spectra (co-spectra uv), the VLSMs provide about 60% to Reynolds shear stress in the logarithmic region. This high contribution shows the evidence of VLSMs and how they play an important role in the outer layer. The results also proved that VLSMs are the major contribution to total kinetic energy and Reynolds shear stress in the logarithmic and outer layer.

8.4 Proper Orthogonal Decomposition Analysis

An application of a new POD method is performed in turbulent pipe flow to study the contribution of large scale structures to turbulent statistics in the first POD mode. Major findings are summarized as follow:

- The symmetric pattern of Gaussian distribution for positive and negative values of POD temporal coefficient a_1 indicates that the first POD mode carries the same contribution from Reynolds Q_2 and Q_4 events correspondingly with values larger than twice of their normalized RMS values, i.e. $\mid a_1 \mid \geq 2\ \sigma_{a_1}$.

- Features of large structures are their alignment in the stream-wise direction to produce regions of low and high stream-wise momentums in the inner and outer layers.

- Instantaneous velocity vector maps show that large structures inhabit the log and outer regions of the boundary layer at high Reynolds number to be responsible for the ejection and sweep events generation, which are the major contributors to turbulent production and Reynolds shear stress.

- It is realized that the most energetic structures are presented as many large scale Q_2 or Q_4 events, and small vortex are almost prograde at the near wall. Besides, hairpin vortex packet and large scale Q_4 event are identified as the major structures and the most important contributors to the first POD mode. This demonstration is consistent with Wu 2014 and Discetti et al. 2019 observation.

- The contribution of large scale structures to stream-wise Reynolds stress u^2, wall-normal Reynolds stress v^2 and Reynolds shear stress -uv is identified by computing new ensembles of removed fluctuating velocity fields (i.e. $\mid a_1 \mid > 2\ \sigma_{a_1}$, $\mid a_1 \mid > 1.5\ \sigma_{a_1}$ and $\mid a_1 \mid > 1\ \sigma_{a_1}$) and comparing it with the original velocity field ensemble. This method has been used previously by Wu et al. 2014 in the turbulent boundary layer.

- It is observed that the value of Reynolds stress (u^2) is reduced by increasing the number of removable time coefficient ensembles a_1 to attain 6% for the highest ensemble $\mid a_1 \mid > 2\ \sigma_{a_1}$) and 24% for the lowest ensemble ($\mid a_1 \mid > 1\sigma_{a_1}$) in comparison with the original ensemble. A significant adhesion is noticed for all ensembles at $y/R \leq 0.5$.

- No difference in wall-normal Reynolds stress v^2 is observed. The contribution of Q_2 and Q_4 events to Reynolds shear stress -uv are realised to be more influenced toward the wall region at y/R = 0-0.5 where, the majority of Reynolds shear stress is in this region, to acquire 9.3 % deviation for the highest ensemble ($\mid a_1 \mid > 2\ \sigma_{a_1}$) and 20 % for the lowest ensemble ($\mid a_1 \mid > 1\ \sigma_{a_1}$).

8.5 Recommendations for Future Work

Despite the previous literature, the origin and magnitude of the VLSMs and their relationship to LSMs are still not recognized enough. Therefore, proceeding temporal-spatial analysis such as cross-correlation and self-similarity are recommended to define structure convective velocity and to obtain the two-dimensional stream-wise variances.

Long field of view by using multiple cameras and three-dimensional velocity measurement technique PIV are needed to investigate the three-dimensional topography of large and very large scale motions. Additional investigation can be proposed by implementing active disturbance (active flow control device) on the CoLaPipe. Such a device can create synthetic structures in the boundary layer and can help to study their influence on the natural coherent structures.

Bibliography

[1] Abe H., Kawamura H., and Choi H., 2004, Very large scale structures and their effects on the wall shear-stress fluctuations in a turbulent channel flow up to Re_τ=640, J. Fluids Eng., vol. 126, No. 5, pp. 835-843.

[2] Adrian R and Yao S.C., 1983, Development of Pulsed Laser velocimetry (PLV) for Measurement of Turbulent flow, Univ of Missouri-Rolla, pp. 170-186.

[3] Adrian R.J., 2007, Hairpin vortex organization in wall turbulence.Phys. Fluids 19:04130.

[4] Adrian R.J., Meinhart C.D., and Tomkins C.D., 2000, Vortex organization in the outer region of the turbulent boundary layer., J. Fluid Mech. 422, 1-54.

[5] Adrian R.J. and Marusic I, 2012, Coherent structures in flow over hydraulic engineering surfaces, Journal of Hydraulic Research, vol. 50, No. 5, pp. 451-464

[6] Antoranz A., Ianiroa A. Floresa O., García-Villalbaa M., 2018, Extended proper orthogonal decomposition of non-homogeneous thermal fields in a turbulent pipe flow, Inter. Journal of Heat and Mass Transfer Vol. 118, pp. 1264-1275.

[7] Afzal N., 1976, Millikan's argument at moderately large Reynolds numbers, Phys. Fluids, vol. 19, pp. 600-602.

[8] Afzal N. and Aligarh, 1982, Fully Developed Turbulent Flow in a Pipe: An Intermediate Layer, Ingenieur-Archiv 52 355-377.

[9] Afzal N.and Yajni, K., 1973, Analysis of turbulent pipe and channel flows at moderately large Reynolds number, J. Fluid Mech., vol. 61, pp. 23-31.

[10] Ahn J., Lee H. J., Lee J., Kang J., and Jin Sung H., 2015, Direct numerical simulation of a 30R long turbulent pipe flow at Re_τ = 3008, Physics of Fluids, vol. 27, 065110.

[11] Atkinson C., Buchmann N.A, Amili O., Soria J., 2014, On the appropriate filtering of PIV measurements of turbulent shear flows, Experiments in fluids, vol. 55, 1654.

[12] Bailey S.C. C., Smits A. J., 2010, Experimental investigation of the structure of large- and very large-scale motions in turbulent pipe flow, J. Fluid Mech. 651:339-56.

[13] Bailey S. C. C., Vallikivi M., Hultmark M. and Smits A. J., 2014, Estimating the value of von Kármán's constant in turbulent pipe flow, Fluid Mech., vol. 749, pp. 79-98.

[14] Balakumar B.J., Adrian R.J., 2007. Large- and very-large-scale motions in channel and boundary-layer flows. Philos. Trans. R. Soc. Lond. A 365:665-81.

[15] Baltzer J. and Adrian R., 2011, Structure, scaling, and synthesis of proper orthogonal decomposition modes of inhomogeneous turbulence, Phys. Fluids 23, 015107.

[16] Baltzer J.R., Adrian R.J., Wu X., 2013, Structural organization of large and very large scales in turbulent pipe flow simulation, Journal of Fluid Mechanics, vol.720, pp.236-279.

[17] Benedict L.H., Gould R.D., 1996, Towards better uncertainty estimates for turbulence statistics. Exp Fluids 22(2),129-136.

[18] Blackwelder R.F., Kovasznay L.S.G., 1972, Large-scale motion of a turbulent boundary layer during relaminarization, Journal of Fluid Mechanics, vol. 53, pp. 61-83.

[19] Brown G. and Thomas A., 1977, Large structure in a turbulent boundary layer, The Physics of Fluids vol. 20, pp. 243- 253.

[20] Bullock K. J., Cooper R. E. and Abernathy F. H., 1978, Structural similarity in radial correlations and spectra of longitudinal velocity fluctuations in pipe flow., J. Fluid Mech. 88, 585-608.

[21] Bruun H. H., Khan M. A., Al-Kayiem H. H. and Fardad A. A., 1988, Velocity calibration relationships for hot-wire anemometry, J. Phys. E: Sci. Instrum. 21 225-232.

[22] Bruun H. H., 1995, Hot Wire Anemometry: Principles and Signal Analysis (Oxford: Oxford University Press).

[23] Chanetz B., Délery J., Gilliéron P., Gnemmi P., Gowrec E., Perrier P., 2020. Experimental Aerodynamics, Springer Tracts in Mechanical Engineering.

[24] Chen H., Reuss D. L., and Sick V., 2012, On the use and interpretation of proper orthogonal decomposition of in-cylinder engine flows, Meas. Sci. Technol., vol. 23, no. 8, p. 85302.

[25] Chin C., Ng H.C.H., Blackburn H.M., Monty J.P., Ooi A., 2015, Turbulent pipe flow at $Re_\tau \approx 1000$: A comparison of wall-resolved large-eddy simulation, direct numerical simulation and hot-wire experiment, Computers and Fluids, vol. 122, pp.26-33.

[26] Cimbala J. M. and Park W. J., 1990, A direct hot-wire calibration technique to account for ambient temperature drift in incompressible, flow Exp. Fluids vol. 8 pp. 299-300.

[27] Collis D. C., 1952, The dust problem in hot-wire anemometry Aeronaut. Q. 4 pp.93-102.

[28] Comte-Bellot G., 1976, Hot-wire anemometry Annu. Rev. Fluid Mech. Vol.8 pp. 209-31.

[29] Dantec Dynamics, 2018, Probes for Hot-wire Anemometry, accessed on December 2020, <https://www.dantecdynamics.com/components/hot-wire-and-hot-film-probes>, Publication No.: 238-v9.

[30] del Alamo J.C., Jiménez J., 2009, Estimation of turbulent convection velocities and corrections to Taylor's approximation.J. Fluid Mech. 640:5-26.

[31] del Alamo J.C., Jiménez J., Zandonade P., Moser R.D., 2004, Scaling of the energy spectra of turbulent channels., J. Fluid Mech. 500:135-144.

[32] Deshpande R., Monty J., and Marusic I., 2020, A scheme to correct the influence of calibration misalignment for cross-wire probes in turbulent shear flows, Exp. Fluids 61:1-17.

[33] de Silva C. M., Grayson K., Scharnowski S., Kähler C.J., Hutchins N., Marusic I., 2018, Towards fully-resolved PIV measurements in high Reynolds number turbulent boundary layers with DSLR cameras, J Vis vol. 21, pp. 369-379.

[34] Dennis D.J.C., 2015, Coherent structures in wall-bounded turbulence,Anais da Academia Brasileira de Ciencias vol.87, pp. 1161-1193.

[35] Discetti S., Bellani G., Örlü R., Serpieri J., Vila C. S., Raiola M., Zheng X., Mascotelli L., Alessandro Talamelli A., Ianiro A., 2019, Characterization of very-large-scale motions in high-Re pipe flows, Experimental thermal and fluid Science vol. 104.

[36] Dudderar TD, Simpkins PG, 1977, Laser speckle photography in a fluid medium, Nature, vol. 270, pp. 45-47.

[37] Duggleby A., Ball K.S., Paul M.R., Fischer P.F., 2007, Dynamical eigenfunction decomposition of turbulent pipe flow, Journal of Turbulence, vol. 8, no. 43.

[38] Durst F., Noppenberger S., Still M. and Venzke H., 1996, Influence of humidity on hot-wire measurements Meas. Sci. Technol. 7 1517.

[39] Furuichi N, Terao Y, Wada Y, Tsuji Y., 2015, Friction factor and mean velocity profile for pipe flow at high Reynolds numbers. Phys. Fluids vol. 27, 095108.

[40] Ganapathisubramani B., Longmire E. K., Marusic I., 2003, Characteristics of vortex packets in turbulent boundary layers, J. Fluid Mech. 478:35-46.

[41] Ganapathisubramani B., Longmire E. K., Marusic I., Pothos S. 2005, Dual-plane PIV technique to determine the complete velocity gradient tensor in a turbulent boundary layer, Experiments in Fluids, vol. 39, pp. 222-231.

[42] Gauthier, V., Riethmuller, M. L.1988, Application of PIDV to complex flows: Resolution of the directional ambiguity, Particle Image Displacement Velocimetry, von Kármán Institute for Fluid Dynamics Lecture Series 6

[43] Guala M., Hommema S. E., and Adrian R.J., 2006, Large-scale and very-large-scale motions in turbulent pipe flow, Journal of Fluid Mechanics, vol. 554, pp. 521-542.

[44] Güemes A., Sanmiguel Vila C., Örlü R., Vinuesa R., Schlatter P., Ianiro A., Discetti S., 2019, Flow organization in the wake of a rib in a turbulent boundary layer with pressure gradient, Exp. Thermal and fluid mechanics, vol. 108, pp. 115-124.

[45] Hallol Z., Merbold S., Egbers C., 2020, Contribution of Large and Very Large Scale Motions to the Turbulent Kinetic energy in Turbulent Pipe Flow, Proc. Appl. Math. Mech., vol.20, e202000229.

[46] Head M. R. and Bandyopadhyay P., 1981, New aspects of turbulent boundary-layer structure, Journal of Fluid Mechanics, vol. 107, pp. 297- 338.

[47] Hellström L., Sinha A., Smits A., 2011, Visualizing the very-large-scale motions in turbulent pipe flow, Phys. Fluids 23, 011703.

[48] Hellström L., Ganapathisubramani B., Smits A., 2015, The evolution of large-scale motions in turbulent pipe flow, J. Fluid Mech. vol. 779 pp. 701-715.

[49] Hellström L. and Smits A., 2017, Structure identification in pipe flow using proper orthogonal decomposition, Phil. Trans.R. Soc. A375: 20160086.

[50] Holmes P. J., Lumley J. L., Berkooz G., Mattingly J. C., Ralf W. Wittenberg R. W.,1997, Low-Dimensional Models of Coherent Structures in Turbulence, Physics Reports vol. 287 pp. 337-384.

[51] Holmes P. J., Lumley J. L., Berkooz G., Rowley C.W., 2012, Turbulence, coherent structures, dynamical systems and symmetry, Cambridge University Press, second edition.

[52] Hultmark M. and Smits A.J., 2010, Temperature corrections for constant temperature and constant current hot-wire anemometers, Meas. Sci. Technol. 21 105404.

[53] Hultmark M., Vallikivi M., Bailey S. C. C. and Smits A. J., 2013, Logarithmic scaling of turbulence in smooth-and rough-wall pipe flow, J. Fluid Mech. 728,376-395.

[54] Hutchins N., Marusic I., 2007a, Evidence of very long meandering features in the logarithmic region of turbulent boundary layers, J. Fluid Mech. 579:1-2.

[55] Hutchins N., Marusic I., 2007b, Large-scale influences in near-wall turbulence, Philos. Trans. R. Soc. Lond. A, vol. 365, 647-664.

[56] Hutchins N., Nickels T. B., Marusic I., and Chong M. S., 2009, Hot-wire spatial resolution issues in wall-bounded turbulence, Journal of Fluid Mechanics, vol. 635, pp. 103-136.

[57] Hutchins N., Hambleton W. T., and Marusic I., 2005, Inclined cross-stream stereo particle image velocimetry measurements in turbulent boundary layers, J. Fluid Mech., vol.541, pp. 21-54.

[58] Jiménez J. and Moin P., 1991, The minimal flow unit in near-wall turbulence., J. Fluid Mech., vol. 225, pp. 213-240.

[59] Jiménez J., 1998, The largest scales of turbulent wall flows.Center for Turbulence Research, Annual Research Briefs, pp. 137-154. Standford University.

[60] Jin C. and Ma H., 2018, POD Analysis of Entropy Generation in a Laminar Separation Boundary Layer, Energies vol. 11, no. 3003.

[61] Johnstone A., Uddin M., Pollard A., 2005, Calibration of hot-wire probes using non-uniform mean velocity profiles. Exp Fluids; 39:52532.

[62] Kellnerová R., Kukáka L., Juráková K., Uruba V., Jáour Z., 2012, PIV measurement of turbulent flow within a street canyon: Detection of coherent motion, Journal of Wind Engineering and Industrial Aerodynamics vol. 104-106, pp. 302-313.

[63] Kim K.C., Adrian R.J., 1999, Very large-scale motion in the outer layer, Phys. Fluids vol.11, pp. 417-422.

[64] Kompenhans J. and Reichmuth J.,1986, Particle imaging velocimetry in a low turbulent wind tunnel and other low facilities. In: Proceedings of the IEEE Montech.86 conference, Montreal, Canada, October 1986. AGARD conference proceedings no. 399, paper 35.

[65] König F., Zanoun E.S., Öngüner E., and Egbers C., 2014. The CoLaPipe - the new Cottbus Large Pipe test facility at BTU Cottbus. Review Scientific Inst., 85, No. 7, 075115-1075115-9.

[66] König F., 2015. Investigation of high Reynolds number Pipe flow. Doctoral thesis,Brandenburg University of Technology Cottbus-Senftenberg, ISBN-13 978-3-73699-049-4.

[67] Lee T.and Budwig R., 1991, Two improved methods for low-speed hot wire calibration, Meas. Sci. Technol. 2 M3-646.

[68] Lee J.H. and Sung H.J., 2013, Comparison of very-large-scale motions of turbulent pipe and boundary layer simulations, Physics of Fluids, vol. 25, pp.045103.

[69] Lee J., Ahn J. & Sung H. J., 2015, Comparison of large and very-large-scale motions in turbulent pipe and channel flows. Phys. Fluids, vol. 27 (2), 025101.

[70] Lee J., Lee J. H., Choi J. & Sung H. J., 2014, Spatial organization of large and very-large-scale motions in a turbulent channel flow., J. Fluid Mech. vol.749, pp. 818-840.

[71] Li S., Jiang N., Yang S., Huang Y., and Wu Y.,2018, Coherent structures over riblets in turbulent boundary layer studied by combining time-resolved particle image velocimetry (TRPIV), proper orthogonal decomposition (POD), and finite-time Lyapunov exponent (FTLE), Chin. Phys. B, Vol. 27, No. 10, 104701.

[72] Liu Z., Adrian R., and Hanratty T., 2001, Large-scale modes of turbulent channel flow: Transport and structure, J. Fluid Mech. vol. 448, pp. 53-80.

[73] Liu Z.C., Landreth C.C., Adrian R.J., Hanratty T.J., 1991, High resolution measurement of turbulent structure in a channel with particle image velocimetry, Experiments in Fluids, vol. 10, pp. 301-312

[74] Lourenco A., 1984, Sprinkler Device, United States Patent, 4,462,545.

[75] Lozano-Durán A. and Jiménez J., 2014, Effect of the computational domain on direct simulations of turbulent channels up to Re = 4200, Phys. Fluids 26 011702.

[76] Lumley J. L., 1967, The structures of inhomogeneous turbulent flow, in Proceedings of the International Colloquium on the Fine Scale Structure of the Atmosphere and its Influence on Radio Wave Propagation, pp. 166-178.

[77] Mckeon B.J., 2017, The engine behind (wall) turbulence: perspectives on scale interactions, J. Fluid Mech., vol. 817, P1.

[78] Martins F.J.W.A., Foucaut J. M., Stanislas M., and Azevedo L. F. A, 2019, Characterization of near-wall structures in the log-region of a turbulent boundary layer by means of conditional statistics of tomographic PIV data, Exp. Thermal and fluid science, vol. 105, pp. 191-205.

[79] Marusic I, Monty JP, Hultmark M, Smits AJ. 2013 On the logarithmic region in wall turbulence, J. Fluid Mech.716, R3.

[80] Marusic I. and Kunkel G., 2003, Streamwise turbulence intensity formulation for flat-plate boundary layers, Physics of Fluids, vol. 15,no.8,pp. 2461-2462.

[81] Marusic I., Mathis R., and Hutchins N., 2010, Predictive model for wall-bounded turbulent flow.Science, vol.29, pp. 193-196.

[82] Mathis R., Monty JP., Hutchins N., and Marusic I., 2009, Comparison of large-scale amplitude modulation in turbulent boundary layers, pipes, and channel flows, Physics of Fluids, vol. 21; 111703.

[83] Meyers J.F.,1991, Generation of Particles and Seeding, VKI, Laser Velocimetry, Volume 1, pp. 1-45.

[84] Meynart R, 1983 a, Instantaneous velocity field measurements in unsteady gas flow by speckle velocimetry, Applied optics, vol. 22, Issue 4, pp. 535-540, doi:10.1364/AO.22.000535

[85] Meynart R, 1983 b, Speckle velocimetry study of vortex pairing in a low Re unexcited jet, The physics of fluids, vol. 26, pp. 2074-2079 doi: 10.1063/1.864411.

[86] Morrison J. F., McKeon B. J., Jiang W. and Smits A., 2004, Scaling of the streamwise velocity component in turbulent pipe flow, J. Fluid Mech. vol. 508, pp. 99-131.

[87] Monty J.P., Stewart J.A, Williams R.C., Chong M.S., 2007, Large-scale features in turbulent pipe and channel flows., J. Fluid Mech. 589:147-56.

[88] Monty J.P., Hutchins N., Ng H.C.H., Marusic I., Chong M.S., 2009, A comparison of turbulent pipe, channel and boundary layer flows, J. Fluid Mech. 632:431-42.

[89] Munir S., Siddiqui M. I., Heikal M., Aziz A. R. A., and de Sercey G., 2015, Identification of dominant structures and their flow dynamics in the turbulent two-phase flow using POD technique, J. Mech. Sci. Technol., vol. 29, no. 11, pp. 4701-4710.

[90] Nagib H.M., Chauhan K.A., Monkewitz P.A., 2007, Approach to an asymptotic state for zero pressure gradient turbulent boundary layers, Philos. Trans. R. Soc. Lond. A, vol. 365, pp. 755-770.

[91] Nekrasov Y. P. and Savostenko P. I., 1991, Pressure dependence of hot-wire anemometer readings, Meas. Technol. 34 462-5.

[92] Ng H.C.H., Monty J. P., Hutchins N., Chong M. S., Marusic I., 2011. Comparison of turbulent channel and pipe flows with varying Reynolds number. Exp Fluids. 51,1261-1281.

[93] Nickels T.B., Marusic I., Hafez S.M., Chong M.S., 2005, Evidence of the k^{-1} law in a high-Reynolds-number turbulent boundary layer, Phys. Rev. Lett. 95:074501.

[94] Öngüner E., Zanoun E.-S., Fiorini T., Bellani G., Shahirpour A., Egbers Ch., Talamelli A., 2017a, Wavenumber Dependence of Very Large-Scale Motions in CICLoPE at $4800 \leq Re_\tau \leq 37.000$, in: Örlü R., Talamelli A., Oberlack M., Peinke J. (eds) Progress in Turbulence VII. Springer Proceedings in Physics, vol 196. Springer, pp 101-106, ISBN 978-3-319-57933-7.

[95] Öngüner E., Zanoun E.-S., Egbers Ch., 2017b, Streamwise Auto-Correlation Analysis in Turbulent Pipe Flow Using Particle Image Velocimetry at High Reynolds Numbers, in: Örlü R., Talamelli A., Oberlack M., Peinke J. (eds) Progress in Turbulence VII. Springer Proceedings in Physics, vol 196. Springer, pp 107-112, ISBN 978-3-319-57933-7

[96] Öngüner E., 2018, Experiments in Pipe Flows at Transitional and Very High Reynolds Numbers, PhD thesis, Brandenburg University of Technology Cottbus-Senftenberg, ISBN-13-978-3-73699-783-7.

[97] Österlund J.M. and Johansson A. V.,1999, Turbulence Statistics of Zero Pressure Gradient Turbulent Boundary Layers, Tech. rep. Sweden: KTH Mechanics.

[98] Örlü R, Alfredsson PH., 2012, Comment on the scaling of the near-wall streamwise variance peak in turbulent pipe flows.Exp. Fluids54, 1431.

[99] Panton R.L., 1997, A Reynolds stress function for wall layers, J. Fluids Eng., vol. 119, pp.325-330.

[100] Patel V.C., Head M.R., 1969. Some observations on skin friction and velocity profiles in fully developed pipe and channel flow. J Fluid Mech. 38: 181-201.

[101] Penga C., Genevaa N., Guob Z., Wang L., 2018, Direct numerical simulation of turbulent pipe flow using the lattice Boltzmann method, Journal of Computational Physics 357 16-42.

[102] Perry A.E., 1982, Hot Wire Anemometry (Oxford: Oxford University Press).

[103] Perry A.E. and Abell C. J., 1975, Scaling laws for pipe-flow turbulence., J. Fluid Mech.67, 257-271.

[104] Perry A.E., Henbest S. and Chong M. S., 1986, A theoretical and experimental study of wall turbulence., J. Fluid Mech. 165, 163-199.

[105] Perry A.E. and Marusic I., 1995, A wall-wake model for the turbulence structure of boundary layers. Part 1. Extension of the attached eddy hypothesis. J. Fluid Mech., vol. 298, pp.361-88.

[106] Raffel M., Willert C.E., Wereley S.T., Kompenhans J.,2007, Particle Image Velocimetry, Springer, second edition.

[107] Reynolds O., 1883, An experimental investigation of the circumstances which determine whether the motion of water shall be direct or sinuous, and of the law of resistance in parallel channels, Philos. Trans., vol. 174, pp. 935-982.

[108] Robinson S.K., 1991, Coherent motions in turbulent boundary layers., Annu. Rev. Fluid Mech., vol. 23, pp.601-39.

[109] Rott N., 1990, Note on the History of the Reynolds Number, Annu. Rev. Fluid Mech., vol. 22, pp. 1-11.

[110] Saxton-Fox T. and McKeon B.J., 2017, Coherent structures, uniform momentum zones and the streamwise energy spectrum in wall-bounded turbulent flows, s.J. Fluid Mech., vol.826, pp. R6.

[111] Schlatter P., Örlü R., Li Q., Brethouwer G., Fransson J. H. M., Johansson A. V., Alfredsson P. H., and Henningson D. S., 2009, Turbulent boundary layers up to Re_θ=2500 studied through simulation and experiment, Physics of Fluids, vol. 21, pp.051702.

[112] Sirovich L., 1987, Turbulence and the dynamics of coherent structures. I. coherent structures, Quart. Appl. Math. vol. 45 pp. 561-571.

[113] Selvam, K., Öngüner, E., Peixinho, J., Zanoun, E.-S., Egbers, C., 2018. Wall Pressure in developing turbulent pipe flows. J. Fluids Eng. ASME, Vol. 140 / 081203-1, 2018.

[114] Sreenivasan K.R., Sahay A., 1997, The persistence of viscous effects in the overlap region and the mean velocity in turbulent pipe and channel flows. (ed. R. Panton). pp. 253-272. Computational Mechanics Publication.

[115] Sesterhenn J., Shahirpour A., 2019, A characteristic dynamic mode decomposition, Theor. Comput. Fluid Dyn. pp 1-25, ISSN: 0935-4964.

[116] Srinath S., 2017, Modeling and prediction of near wall turbulent flows. Fluids mechanics. Ecole Centrale de Lille, PhD thesis.

[117] Smits A.J., McKeon B.J., and Marusic I., 2011, High-Reynolds number wall turbulence, Annu. Rev. Fluid Mech. 2011. 43:353-75.

[118] Talluru k.M., Kulandaivelu V., N Hutchins N.and Marusic I., 2014, A calibration technique to correct sensor drift issues in hot-wire anemometry, Meas. Sci. Technol., vol. 25, pp. 105304.

[119] Tang Z. and Jiang N., 2012, Dynamic mode decomposition of hairpin vortices generated by a hemisphere protuberance, Sci. China Phys., Mech. Astron., vol. 55, no. 1, pp. 118-124.

[120] Taylor G.I., 1938, The spectrum of turbulence.Proc. R. Soc. Lond.164:476-90.

[121] Tennekes, H., 1968, Outline of a second-order theory of turbulent pipe flow. A.I.A.A.J., vol. 6, pp. 1735-1740.

[122] Tennekes H., Lumley JL., 1972, Spectral dynamics, in "A First Course in Turbulence", The M.I.T. Press, Cambridge, MA, pp. 248-287.

[123] Theodorsen T., 1952, Mechanism of turbulence., InProc. 2nd Midwest. Conf. Fluid Mech., Mar. 17-19, pp. 1-19.Columbus: Ohio State Univ.

[124] Tomkins C.D., Adrian R.J., 2003, Spanwise structure and scale growth in turbulent boundary layers., J. Fluid Mech.490:37-74.

[125] Tomkins C.D., Adrian R.J., 2005, Energetic spanwise modes in the logarithmic layer of a turbulent boundary layer, J. Fluid Mech., vol. 545, pp. 141-162.

[126] Tropea C., Yarin A. L. and Foss J. F., 2007, Springer Handbook of Experimental Fluid Mechanics Vol 1 (Berlin: Springer).

[127] Tutkun M., George W.K., Delville J., Stanislas M., Johansson P.B.V., 2009, Two-point correlations in high Reynolds number flat plate turbulent boundary layers., J. Turbul.10:1-23

[128] Vallikivi M., Ganapathisubramani B. and Smits A. J., 2015. Spectral scaling in boundary layers and pipes at very high Reynolds numbers. J. Fluid Mech., vol. 771, pp. 303-326.

[129] Vila C. S., Örlü R., Vinuesa R., Schlatter P., Ianiro A., Discetti S., 2017 Adverse pressure-gradient effects on turbulent boundary layers: statistics and flow-field organization, Flow Turbul. Combust. Vol. 99 pp.589-612.

[130] Wallace J. M., 2016, Quadrant analysis in turbulence research: History and evolution," Annu. Rev. Fluid Mech., vol. 48, pp. 131-158. Willmarth W. W. and Bogar T. J., 1977, Survey and new measurements of turbulent structure near the wall Phys. Fluids, vol. 20, pp. 9-21.

[131] Wang G. and Richter D.H., 2019, Two mechanisms of modulation of very-large-scale motions by inertial particles in open channel flow, J. Fluid Mech., vol. 868, pp.538-559.

[132] Willmarth W. W, and Sharma L. K., 1984, Study of turbulent structure with hot wires smaller than the viscous length J. Fluid Mech. Vol.142 pp.121-49.

[133] Wosnik M., Castillo L., and George W. K.,2000, A theory for turbulent pipe and channel flows. J. Fluid Mech., 421:115-145.

[134] Wu Y., 2008, Experimental investigation of highly irregular roughness effects in wall turbulence, PhD thesis (University of Illinois at Urbana-Champaign).

[135] Wu Y. and Christensen K. T., 2010, Spatial structure of a turbulent boundary layer with irregular surface roughness, J. Fluid Mech. 655, 380-418.

[136] Wu Y., 2014, A study of energetic large-scale structures in turbulent boundary layer, Phys. Fluids 26, 045113.

[137] Wu Y., Tang Z., Yang S., Skote M., Tang H., Zhang G. and Shan Y., 2017, Proper-orthogonal-decomposition study of turbulent near wake of S805 airfoil in deep stall, AIAA J., vol. 55, no. 6, pp. 1959-1969.

[138] Wyatt L. A., 1953, A technique for cleaning hot-wires used in anemometry, J. Sci. Instrum., vol.30, pp. 13.

[139] Yang S., Wang Y., Yang M., Song Y., and Wu Y., 2019, POD-Based Data Mining of Turbulent Flows in Front of and on Top of Smooth and Roughness-Resolved Forward-Facing Steps, IEEE vol.7 pp. 18234-18255.

[140] Yin S., Fan Y., Sandberg M., Li Y., 2019, PIV based POD analysis of coherent structures in flow patterns generated by triple interacting buoyant plumes, Building and Environment, vol. 158, pp. 165-181.

[141] Yue Z., Malmström T. G., 1998, A simple method for low-speed hot-wire anemometer calibration. Meas Sci Technol;9:1506-10.

[142] Zagarola M.V. and Smits A.J., 1998, Mean-flow scaling of turbulent pipe flow, J. Fluid Mech. 373, 33-79.

[143] Zanoun E.-S., Öngüner E., Shahirpour A., Merbold S., Egbers Ch., 2017, Spectral Analysis of Large Scale Structures in Fully Developed Turbulent Pipe Flow, PAMM. Proc. Appl. Math. Mech. 17, 647-650, DOI 10.1002/pamm.201710293.

[144] Zanoun E.-S., Egbers Ch., Örlü R., Fiorini T., Bellani G., Talamelli A., 2019, Experimental evaluation of the mean momentum and kinetic energy balance equations in turbulent pipe flows at high Reynolds number, Journal of Turbulence, DOI: 10.1080/14685248.2019.1628968

[145] Zhang Q., Liu Y., and Wang S., 2014, The identification of coherent structures using proper orthogonal decomposition and dynamic mode decomposition, J. Fluids Struct., vol. 49, pp. 53-72.

[146] Zhou J., Adrian R.J., Balachandar S., Kendall T.M., 1999, Mechanisms for generating coherent packets of hairpin vortices in channel flows., J. Fluid Mech. 387:353-96.

www.ingramcontent.com/pod-product-compliance
Ingram Content Group UK Ltd.
Pitfield, Milton Keynes, MK11 3LW, UK
UKHW022000190726
13853UKWH00004B/1645

9 783736 975019